Practical Numerical C Programming

Finance, Engineering, and Physics Applications

Philip Joyce

Apress®

Practical Numerical C Programming: Finance, Engineering, and Physics Applications

Philip Joyce
Goostrey, UK

ISBN-13 (pbk): 978-1-4842-6127-9 ISBN-13 (electronic): 978-1-4842-6128-6
https://doi.org/10.1007/978-1-4842-6128-6

Managing Director, Apress Media LLC: Welmoed Spahr
Acquisitions Editor: Steve Anglin
Development Editor: Matthew Moodie
Coordinating Editor: Mark Powers

Cover designed by eStudioCalamar

Cover image by Erik Eastman on Unsplash (www.unsplash.com)

Distributed to the book trade worldwide by Apress Media, LLC, 1 New York Plaza, New York, NY 10004, U.S.A. Phone 1-800-SPRINGER, fax (201) 348-4505, e-mail orders-ny@springer-sbm.com, or visit www.springeronline.com. Apress Media, LLC is a California LLC and the sole member (owner) is Springer Science + Business Media Finance Inc (SSBM Finance Inc). SSBM Finance Inc is a **Delaware** corporation.

For information on translations, please e-mail booktranslations@springernature.com; for reprint, paperback, or audio rights, please e-mail bookpermissions@springernature.com.

Apress titles may be purchased in bulk for academic, corporate, or promotional use. eBook versions and licenses are also available for most titles. For more information, reference our Print and eBook Bulk Sales web page at http://www.apress.com/bulk-sales.

Any source code or other supplementary material referenced by the author in this book is available to readers on GitHub via the book's product page, located at www.apress.com/978-1-4842-6127-9. For more detailed information, please visit http://www.apress.com/source-code.

Printed on acid-free paper

Table of Contents

About the Author

Philip Joyce has 28 years of experience as a software engineer – working on control of steel production, control of oil refineries, communications software (pre-Internet), office products (server software), and computer control of airports. This involved programming in Assembler, COBOL, Coral66, C, and C++. Philip was a mentor to new graduates in the company. He also has MSc in computational physics (including augmented matrix techniques and Monte Carlo techniques using Fortran) from Salford University, 1996. Philip is a chartered physicist and member of the Institute of Physics (member of the higher education group). His first book, *Numerical C,* was published by Apress in September 2019.

About the Technical Reviewer

Juturi Narsimha Rao has 9 years of experience as a software developer, lead engineer, project engineer, and individual contributor. His current focus is on advanced supply chain planning between the manufacturing industries and vendors.

Acknowledgments

Thanks to my wife, Anne, for her support, my son Michael, and my daughter Katharine.

Michael uses regression techniques in his work and has shared some ideas with me.

Katharine is a software engineer working for a UK bank. All three have mathematics degrees.

Thanks to everyone on the Apress team who helped me with the publication of this, my second book. Special thanks to Mark Powers, the coordinating editor, for his advice; Steve Anglin, the acquisitions editor; Matthew Moodie, the development editor; and Juturi Narsimha Rao, the technical reviewer.

Introduction

The C programming language is an important language in many computer applications. It is the basis of C++ and C#. This book will demonstrate how to use the C language to solve problems in finance, commercial/industrial systems, and physics.

A basic familiarity with mathematics is assumed along with some experience of the basics of computer programs.

The first chapter reviews the basic areas that C can be used in. A more detailed introduction to C is contained in my *Numerical C* book.

The chapters following this C review are grouped into finance (including regression, CAPM, and asset pricing), commercial applications (supermarket stock control, airport flight information, and power plant control), and various physics applications. The Graph package has been used to display the results of programs.

There are exercises in each chapter with answers and suggested code at the end of the book. The book's source code can be accessed via the **Download Source Code** link located at `www.apress.com/9781484261279`.

CHAPTER 1

Review of C

This chapter reviews the properties of the C programming language. Example programs are given to illustrate the different areas that C covers, for example, for, while, do-while loops, user-defined functions, switches, mathematical functions, file access, and so on.

The programs tend to bring together similar properties, for example, mathematical functions, and incorporate them as single programs. The reader can just use the part of these programs that they need for their program.

1.1 Arithmetic

This program starts with the basic process of asking the user to enter some data. Here, we will use the term "in the location c" to mean "in the address of the variable c in the local stack space."

Firstly, it uses the `printf` command to write to the user's command line to say "Enter character". When the user types in a character, the `getchar` function reads it and places it into the location c. It then tells the user the character they have entered, firstly using `printf` to say "Character entered" and then `putchar` with `c` as the parameter to write the contents of `c` to the command line. In this case the location `c` is a character location denoted by `char`.

If we want to read integers rather than characters, we define the location where it is to be stored as int. In this case we call the `int this_is_a_number1`. Here, we use the more widely used command `scanf` to read in the integer. We specify `this_is_a_number1` as a parameter to the call, and as a first parameter, we specify %d to say that it is an integer.

We can repeat this with another variable `this_is_a_number2`. We can now add these two variables using the coding `total= this_is_a_number1+ this_is_a_number2` where `total` has to be defined as an integer. Again, we can use the `printf` function to display our answer from `total`.

P. Joyce, *Practical Numerical C Programming*, https://doi.org/10.1007/978-1-4842-6128-6_1

We can do similar things with floating point numbers. We define them as `float` rather than `int`. We can subtract numbers using - rather than +. Similarly, we can multiply using * and divide using /.

The following is the code for our arithmetic calculations:

```
/* ch1arith.c */
/* Read, display, and arithmetic */
/* Read input data from the command line */
/* and read it into the program. */
/* Also write the data back to the */
/* command line. Basic arithmetic */
/* done on input data. */
#define _CRT_SECURE_NO_WARNINGS

#include <stdio.h>
int main ()
{
      char c;     /* Declared character variable */
      int this_is_a_number1, this_is_a_number2, total;     /* Declared integer
                                                              variables */
      float float_number1, float_number2,float_total;       /* Declared float
                                                              variables*/

/* Read and display a character */

      printf("Enter character: "); /* Tell the user to enter a character */
      c = getchar(); /* Read the character in and store in c */

      printf("Character entered: "); /* Tell the user what was entered */
      putchar(c); /* Write the char into variable c */

/* Read in two integers, add them, and display the answer */

      printf("\nPlease enter an integer number:\n ");
      scanf("%d", &this_is_a_number1); /* Read number into this_is_a_number1 */
      printf("You entered %d\n", this_is_a_number1);
```

```
    printf("Please enter another integer number: \n");
    scanf("%d", &this_is_a_number2); /* Read number into this_is_a_number2 */
    printf("You entered %d\n", this_is_a_number2);

    total = this_is_a_number1 + this_is_a_number2; /* Add two numbers
                                                    store in total */
    printf("sum of your two integer numbers is %d\n", total);
    /* Write result to command line */

/*  Add two floating point numbers */

    printf("Please enter a decimal number:\n ");
    scanf("%f", &float_number1); /* Read decimal number into float_
                                 number1  */
    printf("You entered %f\n", float_number1);

    printf("Please enter another decimal number: \n");
    scanf("%f", & float_number2); /*Read decimal number into float_
                                  number2 */
    printf("You entered %f\n", float_number2);

    float_total = float_number1+float_number2; /* Add the numbers */
    printf("sum of your two decimal numbers is %f\n", float_total);
    /* Write result to command line */

/* Multiply two floating point numbers */

    float_total = float_number1 * float_number2; /* Multiply the numbers */
    printf("product of your two decimal numbers is %f\n", float_total);
    /* Write result to command line */

/* Divide two floating point numbers */

    /* Divide the numbers */
    /* Place answer into float_total */

    float_total = float_number1 / float_number2);
```

```
        /* Write result to command line */

        printf("quotient of your two decimal numbers is %f\n", float_total);
        return 0;
}
```

1.2 Switches

A switch statement is a multiway branch statement. A program can perform separate different functions. In order to select which one is required, the program asks the user to select a value, for example, 1 to use the cosine function, 2 to use the sine function, and so on. The program then uses this number in the `switch` command to jump to the relevant code.

This sequence of code is shown as follows:

```
printf("\nPlease enter a character a,b,c,d or e:\n ");
scanf("%c", &this_is_a_character);/* read into this_is_a_character */

switch (this_is_a_character)
{
        case 'a':
                printf("Case1: Value is: %c\n", this_is_a_character);
                break;
```

We can switch on numbers or characters. So, for example, we could ask the user to enter a number from 1 to 5 or a letter from a to e. For characters we read their value using `scanf` with `%c` as a parameter. In the program, if you select a, then the switch jumps to `case` a in the code. In the code here, we print out the fact that we have jumped to case a, but this is only to demonstrate how it works. After the relevant code in case a, the program issues a `break` which jumps to the end of the `switch` options.

If the user is asked to type a to e but they type in f, then the switch goes to the default case. Here, we can just output an error message to the user.

The following code demonstrates switches:

```
/* ch1sw.c */
/* Demonstrate switch case functionality by using switch case */
/* parameter choice as either characters or numbers */
#define _CRT_SECURE_NO_WARNINGS
```

```
#include <stdio.h>

/* Example of a switch operation */
int main()
{
   int this_is_a_number; /* Store area to hold number entered */
   char this_is_a_character; /* Store area to hold character entered */

   printf("\nPlease enter a character a,b,c,d or e:\n ");
   scanf("%c", &this_is_a_character); /* Read into this_is_a_character */

   /* Switch to the specific "case" for the character entered */
   /* then print which switch case was entered */
   switch (this_is_a_character)
   {

         case 'a':
               printf("Case1: Value is: %c\n", this_is_a_character);
               break;
         case 'b':
               printf("Case2: Value is: %c\n", this_is_a_character);
               break;
         case 'c':
               printf("Case3: Value is: %c\n", this_is_a_character);
               break;
         case 'd':
               printf("Case4: Value is: %c\n", this_is_a_character);
               break;
         case 'e':
               printf("Case5: Value is: %c", this_is_a_character);
               break;
         default:
               /* The character entered was not between a, b, c, d, or e */
               printf("Error Value is: %c\n", this_is_a_character);
         }
```

```
        printf("Please enter an integer between 1 and 5:\n ");
        scanf("%d", &this_is_a_number);

        /* Switch to the specific "case" for the number entered */
        /* then print which switch case was entered */
        switch (this_is_a_number)
        {

        case 1:
               printf("Case1: Value is: %d\n", this_is_a_number);
               break;
        case 2:
               printf("Case2: Value is: %d\n", this_is_a_number);
               break;
        case 3:
               printf("Case3: Value is: %d\n", this_is_a_number);
               break;
        case 4:
               printf("Case4: Value is: %d\n", this_is_a_number);
               break;
        case 5:
               printf("Case5: Value is: %d\n", this_is_a_number);
               break;
        default:
               /* The number entered was not between 1 and 5 */
               printf("Error Value is: %d", this_is_a_number);
        }

        return 0;
}
```

1.3 Arrays

As well as defining storage locations as single int, char, or float, we can have a number of separate values contained in the same named location. The locations are called arrays. The following program shows an array of 8 integers defined as int arr1[8] where arr1 is the name we use in our program for this location.

We could now store 8 integers, for example, 53614673 in the array. So here, `arr1[0]` contains 5, `arr1[1]` contains 3, `arr1[2]` contains 6, and so on. Note that we count from 0.

We can, as before, ask the user to enter data, but rather than have 8 sets of `printf` and `scanf` commands, we can use a `forloop`, where we tell the program to perform the same instructions 8 times. We use the storage location i to move from `arr1[0]` to `arr1[1]` and so on, and we also use the `i` location to keep count of how many times to go round the loop. In the for instruction `for(i=0;i<8;i++)`, the `i=0` part sets the count i to 0, the `i++` adds 1 each time we loop, and `i<8` limits the number of times to 8 (note again that we count from 0).

We can also have 2D arrays which are a bit like 2D matrices. We can define an array as `arr2[3][5]` so we could store the matrix. The matrix has 3 rows and 5 columns.

$$\begin{pmatrix} 2 & 3 & 6 & 5 & 10 \\ 4 & 12 & 7 & 8 & 11 \\ 9 & 0 & 12 & 13 & 14 \end{pmatrix}$$

```
in our array as arr2[0][0] = 2 arr2[0][1] = 3 arr2[0][2] = 6 arr2[0][3] = 5
arr2[0][4]=10
arr2[1][0] = 4 arr2[1][1] = 12 arr2[1][2] = 7 arr2[1][3] = 8 arr2[1][4]=11
arr2[2][0] = 9 arr2[2][1] = 0 arr2[2][2] = 12 arr2[2][3] = 13 arr2[2][4]=14
```

Note, again, that we count from 0.

The program asks you to enter the 2D matrix. If you enter a 3x5 matrix, you can enter the data here. The program prints your array at the end.

The code is shown as follows:

```
/* ch1arr.c */
/* Array use and nested forloops */

#define _CRT_SECURE_NO_WARNINGS
```

```
#include <stdio.h>

/* Program to show array use */

int main()

{
   int arr1[8]; /* Define an array of 8 integers */

   int arr2[3][5]; /* 2D array of integers 3 rows and 5 columns*/

   int i, j, k, l;

   /* arr1 1D array */
   /* Ask the user to enter the data */
   printf("enter 8 integer numbers\n");

   for (i = 0;i < 8;i++)
   {
         /* Read the data into array arr1 */
         scanf("%d", &arr1[i]); /* Read into arr1[i] */
   }
   printf("Your 8 numbers are \n");

   for (i = 0;i < 8;i++)
   {
         printf("%d ", arr1[i]); /* Write contents of arr1 to command line */
   }
   printf("\n");

   /* arr2 2D array */

   /* Ask the user to enter the data */
   printf("enter number of rows and columns (max 3 rows max 5 columns) \n");
   scanf("%d %d", &k, &l);
   if (k > 3 || l > 5)
   {
         /* User tried to enter more than 3 rows or 5 columns */
         printf("error - max of 8 for rows or columns\n");

   }
```

```
    else
    {
            printf("enter array\n");
            /* Read i rows and j columns using nested forloop */
            for (i = 0;i < k;i++)
            {
                    for (j = 0;j < l;j++)
                    {
                            /* Read the data into array arr2 */
                            scanf("%d", &arr2[i][j]);
                    }
            }
            printf("Your array is \n");
            /* Print entered 2D array using nested forloop */
            for (i = 0;i < k;i++)
            {
                    for (j = 0;j < l;j++)
                    {
                            printf("%d ", arr2[i][j]);
                    }
                    printf("\n");

                    }
            }
}
```

1.4 Strings

The next program shows the use of string manipulation. Strings are char arrays in the program. Our array "select" is preset with values 's' 'e' 'l' 'e' 'c' 't' '\0'. This is preset this way to show how the characters are stored. We would normally define it as char select[7] = "select";. The second and third arrays are string1 and string2 and preset as shown.

Our first function is strlen which just returns the length of the string you have entered. Here, it returns the length of int data point called len. We can then print this to the user using printf.

The second string function copies one string into the other. So here, we say strcpy(string3,string1) copies the contents of string1 into string3. Again, we can print this out using printf.

Our next function, strcmp, compares two strings. If they are the same, it replies 0.

Our final function concatenates one string onto the end of the other. So here, it concatenates string2 onto string1 giving "This is string1. This is string2".

The code is as follows:

```
/* ch1strings.c */
/* Demonstrate strings */

#define _CRT_SECURE_NO_WARNINGS
#include <stdio.h>
#include <string.h>
/* Program to demonstrate string operations strlen, strcpy, strcat, strcmp */

int main() {
       char select[7] = { 's', 'e', 'l', 'e', 'c', 't','\0' };
       char string1[32] = "This is string1";
       char string2[16] = "This is string2";
       char string3[16];

       int  len;

       /* Print out the lengths of the strings */
       /* strlen returns length of string */

       len = strlen(string1);
       printf("strlen(string1) :  %d\n", len);
       len = strlen(string2);
       printf("strlen(string2) :  %d\n", len);
       len = strlen(string3);
       printf("strlen(string3) :  %d\n", len);

       /* copy string1 into string3 */
       /* strcpy copies string1 into string3 */

       strcpy(string3, string1);
       printf("strcpy( string3, string1) :  %s\n", string3);
```

```
        len = strlen(string3);
        printf("strlen(string3) after copy of string1 into string3 :
        %d\n", len);

        /* strcmp compares strings & returns 0 if they are equal */
        /* strcmp returns negative value if they are not equal */
        /* Compare string1 and string3 (these should be the same)*/

        if (strcmp(string1, string3) == 0)
           printf("strings are the same\n");

        /* Concatenates string1 and string2 */

        strcat(string1, string2);
        printf("strcat( string1, string2):   %s\n", string1);

        /* Total length of string1 after concatenation */
        len = strlen(string1);
        printf("strlen(string1) after cat of string2 onto string1 :
        %d\n", len);
        printf("String as predefined quoted chars: %s\n", select);

        return 0;
}
```

1.5 Mathematical Functions

The commonly used mathematical functions are available for you to call from your program. You need to include the library file `<math.h>` in your program to access these. The first three functions are `cos`, `sin`, and `tan`. Here, you are expected to enter the angle in degrees, but you can change the code if you want so that you can enter the angle in radians. The next three are `arccos`, `arcsin`, and `arctan`. Again, these functions will return the angle in degrees, but again you can change this to radians.

The next three functions are `pow`, `sqrt`, and `fabs`. The `pow` function finds the power of a number. Its first parameter is the number you want to find the power of and the second parameter is the power you want to use. The functions return to answer to the store location answer which is defined as a `double float`. The function `sqrt` finds the square root of the number you enter as its parameter. Again, it returns the answer into `answer`. The final function is `fabs`. You supply a number as the parameter and it returns the absolute value of the number.

The code is as follows:

```
/* ch1math.c */
/* Demonstrate mathematics functions */
#define _CRT_SECURE_NO_WARNINGS
#include <stdio.h>

#include <math.h>
#define PI 3.14159265
/* Illustration of the common trigonometric functions */
/* also exponent, natural log, log to base 10 */
/* power, square root, and find absolute value */

int main()
{

       double angle, radianno, answer;

       double  arccos, arcsin, arctan;
       double  expno, natlog, lb10;
       double  pownum, power, sqroot, fabsno;

       /* The cosine function */

       printf("cosine function:\n ");
       printf("Please enter angle in degrees:\n ");
       scanf("%lf", &angle);
       printf("You entered %lf\n", angle);
       radianno = angle * (2 * PI / 360);
       answer = cos(radianno); /* returns cos value to answer */
       printf("cos of %lf is %lf\n", angle, answer);

       /* The sine function */

       printf("sine function:\n ");
       printf("Please enter angle in degrees:\n ");
       scanf("%lf", &angle);
       printf("You entered %lf\n", angle);
       radianno = angle * (2 * PI / 360);
       answer = sin(radianno); /* Returns sin value to answer */
       printf("sin of %lf is %lf\n", angle, answer);
```

```
/* The tangent function */

printf("tangent function:\n ");
printf("Please enter angle in degrees:\n ");
scanf("%lf", &angle);
printf("You entered %lf\n", angle);
radianno = angle * (2 * PI / 360);
answer = tan(radianno); /* Returns tan value to answer */
printf("tan of %lf is %lf\n", angle, answer);

/* The arccos function */

printf("arccos function:\n ");
printf("Please enter arccos:\n ");
scanf("%lf", &arccos);
printf("You entered %lf\n", arccos);
radianno = acos(arccos); /* Returns arccos value to radianno
                            (in radians) */
answer = radianno * (360 / (2 * PI));
printf("arccos of %lf in degrees is %lf\n", arccos, answer);

/* The arcsin function */

printf("arcsin function:\n ");
printf("Please enter arcsin:\n ");
scanf("%lf", &arcsin);
printf("You entered %lf\n", arcsin);
radianno = asin(arcsin); /* Returns arcsin value to radianno
                            (in radians) */

answer = radianno * (360 / (2 * PI));
printf("arcsin of %lf in degrees is %lf\n", arcsin, answer);

/* The arctan function */

printf("arctan function:\n ");
printf("Please enter arctan:\n ");
scanf("%lf", &arctan);
printf("You entered %lf\n", arctan);
```

```
radianno = atan(arctan); /* Returns arctan value to radianno
                          (in radians) */
answer = radianno * (360 / (2 * PI));
printf("arctan of %lf in degrees is %lf\n", arctan, answer);

/* Showing use of exp, log, and log10 functions */
/* Find exponent of entered number */

printf("exponential function:\n ");
printf("Please enter number:\n ");
scanf("%lf", &expno);
printf("You entered %lf\n", expno);

answer = exp(expno);/* returns exponent value to answer */
printf("exponent of %lf is %lf\n", expno, answer);

/* Find natural logarithm of entered number */

printf("natural logarithm function:\n ");
printf("Please enter number:\n ");
scanf("%lf", &natlog);
printf("You entered %lf\n", natlog);
answer = log(natlog); /* Returns natural log value to answer */
printf("natural logarithm of %lf is %lf\n", natlog, answer);

/* find log to base 10 of entered number */

printf("log to base 10 function:\n ");
printf("Please enter number:\n ");
scanf("%lf", &lb10);
printf("You entered %lf\n", lb10);
answer = log10(lb10); /* Returns log to base 10 value to answer */

printf("log to base 10 of %lf is %lf\n", lb10, answer);

/* Showing use of pow, sqrt, and fabs functions */
/* Find x raised to power y number */
```

```
        printf("power:\n ");
        printf("Please enter number:\n ");
        scanf("%lf", &pownum);
        printf("You entered %lf\n", pownum);
        printf("Please enter power:\n ");
        scanf("%lf", &power);
        printf("You entered %lf\n", power);

        answer = pow(pownum, power); /* Returns power of pownum value to
                                         answer */
        printf("%lf raised to power %lf is %lf\n", pownum, power, answer);

        /* Find square root of number */

        printf("square root:\n ");
        printf("Please enter number:\n ");
        scanf("%lf", &sqroot);
        printf("You entered %lf\n", sqroot);

        answer = sqrt(sqroot); /* returns square root of sqroot value to
                                   answer */
        printf("The square root of %lf is %lf\n", sqroot, answer);

        /* Find absolute value of number */

        printf("absolute value:\n ");
        printf("Please enter number:\n ");
        scanf("%lf", &fabsno);
        printf("You entered %lf\n", fabsno);

        answer = fabs(fabsno); /* Returns absolute value of fabsno to answer */
        printf("The absolute value of %lf is %lf\n", fabsno, answer);

        return 0;

}
```

1.6 User-Written Functions

If the same process is to be used several times in a program, then rather than write it out several times, we can define a function. This is a separate piece of code which we can call each time we want to use it. We demonstrate this in our function program. Here, as a simple example, what we want to do several times is to find which of two numbers is greater. We do the comparison in `myfunction` and call this from the main program. We supply the function with two parameters which are the numbers we want to compare. In the program we have defined three numbers, first, second, and third. The following is the call to the function to compare `first` with `second`:

```
myfunction(first , second);
```

We then call `myfunction` with different combinations of these numbers. `myfunction` compares these and prints out the answer.

```
/* ch1func.c */
/* Demonstrate function */
#define _CRT_SECURE_NO_WARNINGS

#include <stdio.h>

/* This code demonstrates what a function does */
/* The function here compares two numbers and says which is bigger */
/* The user enters three numbers and gets told which is bigger than which !*/

       void myfunction(int a,int b); /* Declaration of your function and
                                        its parameters */

       int first , second, third; /* Data fields to hold entered data */
main()
{
       /* Ask the user to enter the numbers to be compared */
       printf( "Please enter first integer number: " );
       scanf( "%d", &first );
       printf( "Please enter second integer number: " );
       scanf( "%d", &second );
       printf( "Please enter third integer number: " );
       scanf( "%d", &third );
```

```
        /* Call "myfunction" with different parameters for each call */
        myfunction(first , second);
        myfunction(first , third);
        myfunction(second , third);
}
void myfunction(int a,int b)
/* The function is outside the main{} part of the program */
/* The function just compares the two parameters, a and b, and says which
is greater*/
{

        if(a>b)
                printf("%d is greater than %d\n", a,b);
        else if (a<b)
                printf("%d is greater than %d\n", b,a);
        else
                printf("%d and %d are equal\n", a,b);
}
```

We can extend the function process so that it returns an answer to the place in the main program where it was called. So here, we call `getmarks` and it returns its answer to the variable `pupil`. The function is supplied with an array containing marks attained in an exam by pupils. The function checks through the array to see which was the highest mark and returns this value. In this case the return value is of variable type `double`. So in the declaration of the function at the start of the program, it has `double` as its prefix. Within the function code in the program, the return variable is called `highest`. This is the value which is returned. The name of the variable in the function does not have to be the same as the variable where the result is placed in the main body of the program (here in the main body the result is placed in the variable pupil).

The following is the code for a function which returns an answer:

```
/* ch1funcans.c */
/* Demonstrate function returning an answer */
#define _CRT_SECURE_NO_WARNINGS
```

```
#include <stdio.h>

/* Function which returns an answer */
/* Finds the pupil in one year of the school with the highest marks */

#include <stdio.h>

/* Declaration of your function and its parameters */
double getmarks(double pupils[]);

int main()
{
        double pupil; /* Declaration of the variable which will contain the
                      answer */

        /* Array with marks for class is preset in the main part of the
        program */
        double marks[] = { 10.6, 23.7, 67.9, 93.0, 64.2, 33.8 ,57.5 ,82.2 ,
        50.7 ,45.7 };

        /* Call function getmarks. The function returns the max marks which
        is then stored in pupil */
        pupil = getmarks(marks);

        /* print the maximum mark */
        printf("Max mark is  = %f", pupil);
        return 0;
}

double getmarks(double pupils[])
{
        /* User-defined function which returns an answer to the call */
        int i;
        double highest;
        highest = 0.0;
        /* Go through all the pupils in turn and store the highest mark */
        for (i = 0; i < 6; ++i)
        {
                if (highest < pupils[i])
                        highest = pupils[i];
```

```
        }
        return highest; /* Returns the value in highest to where the
                           function was called */
}
```

1.7 File Creation

In computing, most data are held in files which are read, acted upon, and then closed. Most files are created in a program and then read by other programs. So here, we will create and write to a file using the program `filecreate`.

The data we want to write to the file is held in a preset structure. The layout of the structure is shown in record. Here, the layout is simple. We will go on to use less simple but more realistic structures later. Here the structure contains a 2D array called `matrix`. We have preset the actual data into the 2D array called `inmat`. Then we will copy the data from `inmat` to `matrix`.

The initial 2D matrix, inmat, is shown in the following with its preset values:

```
double inmat[3][5] = {
            {2.6,3.1,7.4,0.6,9.3},
            {4.9,9.3,0.6,7.4,3.1},
            {8.3,8.8,5.2,2.7,0.8}

    };
```

The following defines the structure called record. In this simple case, record only contains a 2D array called matrix.

```
struct record
    {
        double matrix[3][5];
    };
```

We now define a storage variable called data_record. This variable has the format of struct record. When we access this, we use `data_record.matrix[i][j]`.

```
struct record data_record;
```

First, we open the file `testaug.bin` using `fopen`.

```
ptr = fopen("testaug.bin", "w");
```

The function creates the file "testaug.bin". The second parameter of the call defines what we can do with the file. The main options here are as follows:

r - Open the file to read (create the file if it does not exist).

w - Open the file to write. If it already exists, the contents are deleted first.

a - Open the file to append. It will add to the end of the file (create the file if it does not exist).

r+ - Open the file to read and write.

w+ - Create a file to read and write.

a+ - Open the file to read and append.

Then we write the data to the file using `fwrite`.

```
r1 = fwrite(data_record.matrix, sizeof(data_record.matrix), 1, ptr);
```

The first parameter of `fwrite` is the data, the second parameter is the size of the data, the third parameter is how many records we want to write, and the last parameter is the pointer to the file which we have set up in the `fopen` command. The fwrite command returns a value to the variable r1. The variable r1 contains the number of records written. This can then be printed out to the user.

At the end, we print out what we have written and then close the file using `fclose`.

```
fclose(ptr);
```

Again, we have the pointer to the file as the parameter. This is important if we have two files open in the program and we only want to close one of them.

The program code is shown as follows:

```
/* ch1filecreate.c */
/* filecreate */
/* Creates a file from data preset */
/* into the 2D array of the program */
/* then prints the data to the user */
```

```
#define _CRT_SECURE_NO_WARNINGS
#include<stdio.h>
#include <stdlib.h>
int main()
{
        struct record
        {
                double matrix[3][5];
        };
        int /*counter,*/ i, j;
        FILE *ptr;
        struct record data_record;
        size_t r1;

        double inmat[3][5] = {
                {2.6,3.1,7.4,0.6,9.3},
                {4.9,9.3,0.6,7.4,3.1},
                {8.3,8.8,5.2,2.7,0.8}

        };
        /* Copy preset array to output array */
        for (i = 0;i < 3;i++)
        {
                for (j = 0;j < 5;j++)
                {
                        data_record.matrix[i][j] = inmat[i][j];
                }
        }
        /* Open output file (write/binary) */
        ptr = fopen("testaug.bin", "w");
        if (!ptr)
        {
                /* Error when trying to open file */
                printf("Can not open file");
                return 1; /* quit the program */
        }
```

```
        /* Write output matrix to output file */
        r1 = fwrite(data_record.matrix, sizeof(data_record.matrix), 1, ptr);
        printf("wrote %d elements \n", r1);
        printf("size of data_record.matrix is %d \n", sizeof(data_record.matrix));
        /* Print matrix written to file */
        for (i = 0;i < 3;i++)
        {
                for (j = 0;j < 5;j++)
                {
                        data_record.matrix[i][j] = inmat[i][j];
                        printf("data_record.matrix[%d][%d] = %lf \n", i, j,
                        data_record.matrix[i][j]);
                }
        }

        fclose(ptr); /* Close the file */
        return 0;
}
```

1.8 File Read

We have now created our file `testaug.bin`, so now we write a program to read the data from the file. First, we open the file, as in `filecreate`. Then we read the data from the file using `fread` which has the same parameters as `filecreate`.

```
r1 = fread(data_record.matrix, sizeof(data_record.matrix), 1, ptr);
```

In this case we read the record from the file and place it into data_record.matrix. The variable r1 will contain the number of records read which, again, we can print out to the user.

Then we print out the data we have read and close the file.

The code is as follows:

```
/* ch1fileread.c */
/* fileread */
/* Read the data from a file and write it to command line */
```

```
#define _CRT_SECURE_NO_WARNINGS
#include<stdio.h>
#include <stdlib.h>
        struct record
        {
                double matrix[3][5];
        };
int main()
{
        int counter, i;
        FILE *ptr;
        struct record data_record;
        size_t r1;
        /* Open input file (read/binary) */
        ptr = fopen("testaug.bin", "r");
        if (!ptr)
        {
                /* Error trying to open the file */
                printf("Can not open file");
                return 1;
        }
        /* Read input matrix from input file */
        r1 = fread(data_record.matrix, sizeof(data_record.matrix), 1, ptr);
        printf("read %d elements \n", r1);
        printf("size of struct record is %d \n", sizeof(struct record));
        /* Print matrix read from file */
        /* using nested forloop */
        for (counter = 0; counter < 3; counter++)
        {
                for (i = 0; i < 5; i++)
                {
                        printf("matrix[%d][%d] = %lf \n", counter, i,
                        data_record.matrix[counter][i]);

                }
        }
```

```
        fclose(ptr); /* Close the file */
        return 0;
}
```

1.9 File Create2

This program creates a file which is more typical of those used in the workplace.

This is a file of data for patients under the care of a doctor or hospital. We have a structure of data that we keep for each patient. This consists of their numerical ID, their name, and their blood pressure.

```
struct Patient {
       int PatientID;
       char name[13];
       int BloodPressure;
       };
```

We create a separate structure for each patient.

These are defined in the program as `s10`, `s11`, `s12, and so on` up to `s29`. Data is preset into each structure. The first one is shown as follows:

```
struct Patient s10 = { 10,"Brown ",50 };
```

The output file is opened and we use `fwrite` to write each structure to the file. We can then close the file and open it again and use the `fread` command to read each patient's structure and print it out. If the read command does not return the expected number of records, we can call the function `feof(fp))`. This function returns 1 if an unexpected end of file was read. Otherwise, we call `ferror(fp)` which returns 1 if there is another error.

Finally, we close the file.

The code is as follows:

```
/* ch1filecreate2.c */
/* filecreate2 */
/* Reads from file */
/* Prints out the records sequentially */
/* Finds specific records and prints them */
```

```
#define _CRT_SECURE_NO_WARNINGS
#include<stdio.h>

	/* Structure containing information */
	/* for each record in the file. */
	/* Each record contains data for one patient. */
	/* The data is the patient's ID, name, and blood pressure */
	struct Patient {
	int PatientID;
	char name[13];
	int BloodPressure;
	};
int main()
{
	int i, numread;
	FILE *fp;
	struct Patient s1;
	struct Patient s2;
	/* Preset the data for each patient */
	struct Patient s10 = { 10,"Brown ",50 };
	struct Patient s11 = { 11,"Jones ",51 };
	struct Patient s12 = { 12,"White ",52 };
	struct Patient s13 = { 13,"Green ",53 };
	struct Patient s14 = { 14,"Smith ",54 };
	struct Patient s15 = { 15,"Black ",55 };
	struct Patient s16 = { 16,"Allen ",56 };
	struct Patient s17 = { 17,"Stone ",57 };
	struct Patient s18 = { 18,"Evans ",58 };
	struct Patient s19 = { 19,"Royle ",59 };
	struct Patient s20 = { 20,"Stone ",60 };
	struct Patient s21 = { 21,"Weeks ",61 };
	struct Patient s22 = { 22,"Owens ",62 };
	struct Patient s23 = { 23,"Power ",63 };
	struct Patient s24 = { 24,"Bloom ",63 };
	struct Patient s28 = { 28,"Haver ",68 };
	struct Patient s29 = { 29,"James ",69 };
```

```
/* Open the Patients file */
fp = fopen("patients.bin", "w");

/* Write details of each patient to file*/
/* From the structures defined earlier */
fwrite(&s10, sizeof(s1), 1, fp);
fwrite(&s11, sizeof(s1), 1, fp);
fwrite(&s12, sizeof(s1), 1, fp);
fwrite(&s13, sizeof(s1), 1, fp);
fwrite(&s14, sizeof(s1), 1, fp);
fwrite(&s15, sizeof(s1), 1, fp);
fwrite(&s16, sizeof(s1), 1, fp);
fwrite(&s17, sizeof(s1), 1, fp);
fwrite(&s18, sizeof(s1), 1, fp);
fwrite(&s19, sizeof(s1), 1, fp);
fwrite(&s20, sizeof(s1), 1, fp);
fwrite(&s21, sizeof(s1), 1, fp);
fwrite(&s22, sizeof(s1), 1, fp);
fwrite(&s23, sizeof(s1), 1, fp);
fwrite(&s24, sizeof(s1), 1, fp);
fwrite(&s28, sizeof(s1), 1, fp);
fwrite(&s29, sizeof(s1), 1, fp);

/* Close the file */
fclose(fp);

/* Reopen the file */
fopen("patients.bin", "r");
/* Read and print out all of the records on the file */
for (i = 0;i < 17;i++)
{
       numread = fread(&s2, sizeof(s2), 1, fp);/* read into numread */

       if (numread == 1)
       {
              /* print data for one patient */
              printf("\nPatientID : %d", s2.PatientID);
```

```
            printf("\nName : %s", s2.name);
            printf("\nBloodPressure : %d", s2.BloodPressure);
        }
        else {
            /* If an error occurred on read, then print out message */
            if (feof(fp))
                printf("Error reading patients.bin : unexpected end
                of file fp is %p\n", fp);

            else if (ferror(fp))
                {
                    perror("Error reading patients.bin");
                }
            }
    }
    /* Close the file */
    fclose(fp);
}
```

1.10 File Read2

Our final file program shows how we can move about the file when we are reading it.

We start by opening the file, reading the patient's data, and printing it. We then close the file and reopen it and read the patient's data again but this time only printing out data for the patient with ID=23.

```
fp = fopen("patients.bin", "r");
    for (i = 0;i < 17;i++)
    {
        /* Search the file for patient with ID of 23 */
        fread(&s2, sizeof(s2), 1, fp);
        if (s2.PatientID == 23)
        {
```

```
                /* Found the patient. Print their name */
                printf("\nName : %s", s2.name);
                break;
            }
    }
```

We can then go back to the beginning of the file by using fseek where we have the parameter SEEK_END which moves to the end of the file; then, we can call rewind to go back to the start of the file.

```
fseek(fp, sizeof(s2), SEEK_END);
    rewind(fp);
```

We can then go through the file again. This time, we are finding all of the patients with blood pressure above 63 and printing these out.

Finally, we close the file.

The code is as follows:

```
/* ch1fileread2.c */
/* fileread2 */
/* Reads from file */
/* Reads and prints sequentially */
/* Reads and prints specific records */
#define _CRT_SECURE_NO_WARNINGS
#include<stdio.h>
    struct Patient {
        int PatientID;
        char name[13];
        int BloodPressure;
    };
int main()
{
    FILE *fp;
    struct Patient s2;
    int numread, i;

    /* Open patients file */
    fp = fopen("patients.bin", "r");
```

```
for (i = 0;i < 17;i++)
{
    /* Read each patient data from file sequentially */
    fread(&s2, sizeof(s2), 1, fp);
    /* Print patient ID, name, and blood pressure for each patient */
    printf("\nPatientID : %d", s2.PatientID);
    printf("\n Name : %s", s2.name);
    printf("\nBloodPressure : %d", s2.BloodPressure);
}
fclose(fp);
/* Reopen the patients file */

fp = fopen("patients.bin", "r");
for (i = 0;i < 17;i++)
{
    /* Search the file for patient with ID of 23 */
    fread(&s2, sizeof(s2), 1, fp);
    if (s2.PatientID == 23)
    {
        /* Found the patient. Print their name */
        printf("\nName : %s", s2.name);
        break;
    }
}
/* Go back to the beginning of the file */

fseek(fp, sizeof(s2), SEEK_END);
rewind(fp);
/* Find all patients with blood pressure reading above 63 */

for (i = 0;i < 17;i++)
{
    fread(&s2, sizeof(s2), 1, fp);
    if (s2.BloodPressure > 63)
    {
```

```
                /* Print out name of each patient with blood pressure
                above 63 */

                printf("\nName : %s", s2.name);
            }
    }
    /* Go back to the beginning of the file */

    rewind(fp);
    /* Read and print out the first 3 patients in the file */

    numread = fread(&s2, sizeof(s2), 1, fp);
    if (numread == 1)
    {
            printf("\nPatientID : %d", s2.PatientID);
            printf("\nName : %s", s2.name);
            printf("\nBloodPressure : %d", s2.BloodPressure);
    }
    numread = fread(&s2, sizeof(s2), 1, fp);
    if (numread == 1)
    {
            printf("\nPatientID : %d", s2.PatientID);
            printf("\nName : %s", s2.name);
            printf("\nBloodPressure : %d", s2.BloodPressure);
    }
    numread = fread(&s2, sizeof(s2), 1, fp);
    if (numread == 1)
    {
            printf("\nPatientID : %d", s2.PatientID);
            printf("\nName : %s", s2.name);
            printf("\nBloodPressure : %d", s2.BloodPressure) ;
    }
    /* Close the file */
    fclose(fp);
}
```

1.11 Common Mathematical and Logical Symbols

1. = assign
2. == equals
3. != not equal to
4. < less than
5. > greater than
6. <= less than or equal to
7. >= greater than or equal to
8. && logical AND
9. || logical OR
10. ! logical NOT

EXERCISES

1. Write a program of 3 nested forloops. Each forloop should loop 1000 times. Within the innermost forloop, add 1 to a number. Set the number to zero before you enter the forloops. Print out the number at the end of the program.
2. Write a program to work out the average of 3.1, 0.6, 4.9, 8.7, 0.2, 3.6, 4.9, 7.4, 3.1, and 0.3.
3. Now change your program in question 2 but call a function which returns the answer. Then print it out.
4. Write a program containing the structure for companies.

```
struct company {
      char name[13];/* company name */
      int employees;/* number of employees */
      float yearprofit;/* yearly profit */
      };
```

Open the file and then ask the user to enter the data for each company. Then write this data to the file. Close the file, then reopen it, and read all of the data. Print out the `name`, `employees`, and `yearprofit` for each company.

5. Write a program which reads the data from question 4. Ask the use to enter a value for the profit. Then read the data and print out any company which have a greater profit than the value entered.

6. Write a similar program to question 5 but this time finding the companies with number of employees greater than the entered value.

7. Rewrite the `ch1arith.c` program from this chapter but give the user numbers to enter, one for each function. Then use a switch to jump to that function.

PART I

Finance Applications

Regression

Product Moment Correlation Coefficient

Stock Price Prediction

CHAPTER 2

Regression

Regression is concerned with the relationship between two quantities. For instance, a person's height in relation to their weight or a car's values compared to its age. If we record values of these, we can plot the points on a graph. This type of graph is called a "scatter" graph because the points seem scattered about the graph.

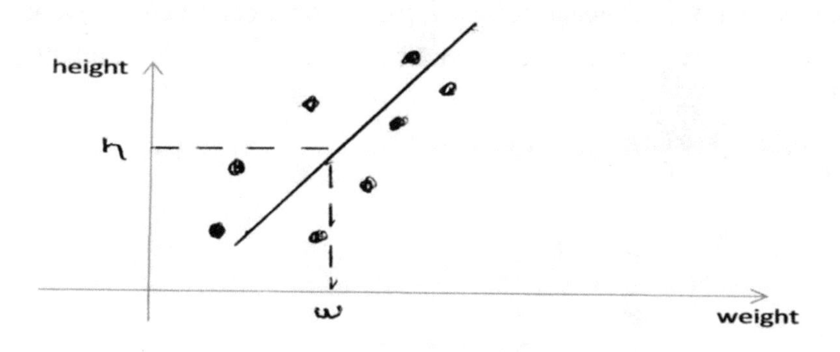

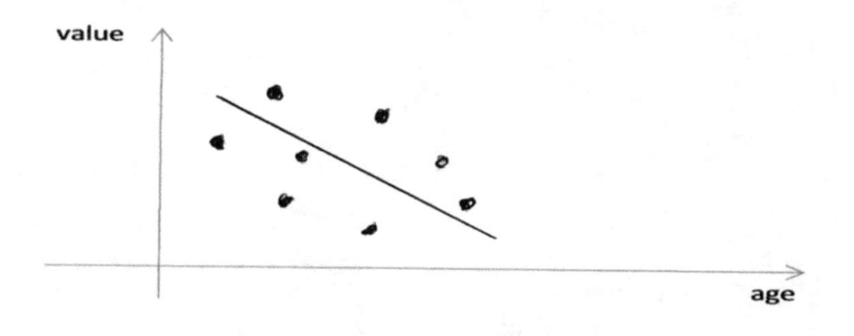

P. Joyce, *Practical Numerical C Programming*, https://doi.org/10.1007/978-1-4842-6128-6_2

The preceding two graphs are scatter graphs. The top graph shows height plotted against weight for different people. The bottom graph shows the value of one particular make and model of a car plotted against the age of the car. Notice that the top graph slopes up from the left showing that the trend is that taller people are heavier than shorter people. The bottom graph slopes down from the left showing that the car's value decreases with age. If there was no relationship between the height of a person and their weight, then the points would be all over the graph and not in the patterns shown. If this is the case, we say that there is no "correlation" between the two quantities. For our case of a slope up from the left where one quantity increases with the other, we call this "positive correlation." For the case of the slope down from the left, we call it "negative correlation."

For the two graphs shown earlier, the points seem to be following a straight line pattern, so we have drawn a straight line "by eye" with approximately the same number of points above the line as below.

We can make this more accurate by using mathematical gauging of what the line should be. This is done by making the distances between the points and our line as small as possible.

There are two possible ways we can do this.

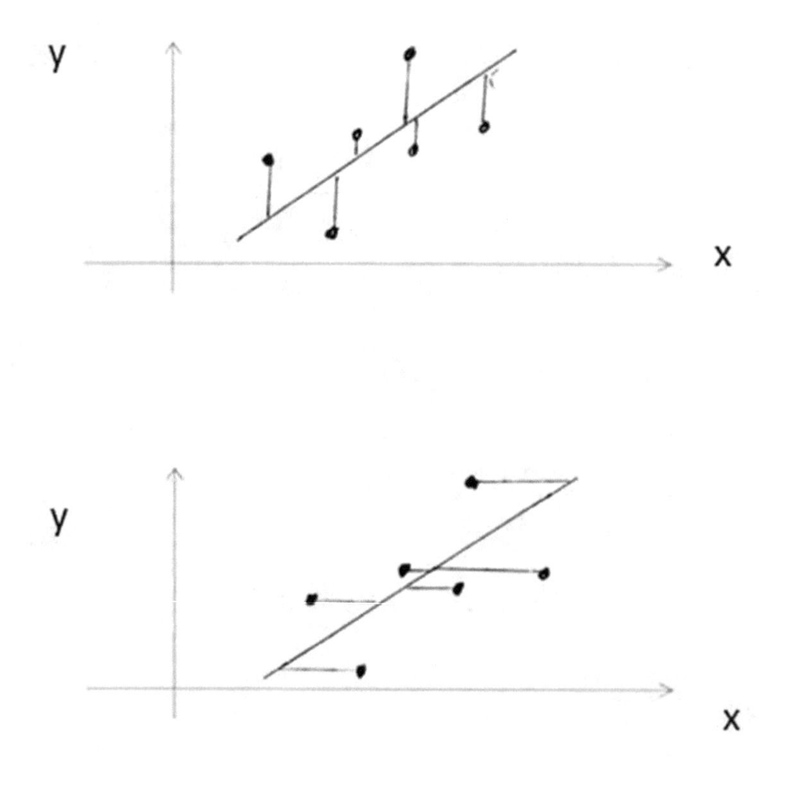

```
/* Store areas for points in the formulas */
float xpoints[12],ypoints[12];
float sigmax,sigmay,sigmaxy,sigmaxsquared,xbar,ybar;
float fltcnt,sxy,sxx,b,a;

int i,points;

fp=fopen("regyonx.dat","w"); /* Create and open the file in "write" mode*/

printf("enter number of points (max 12 ) \n");
scanf("%d", &points);
if(points>12)
{
     printf("error - max of 12 points\n"); /* User entered more
                                                than 12 */

}
else
{
     /* Preset store areas to zero */
     sigmax=0.0;
     sigmay=0.0;
     sigmaxy=0.0;
     sigmaxsquared=0.0;

     /* User enters points */
     for(i=0;i<points;i++)
     {
          printf("enter point (x and y separated by space) \n");
          scanf("%f %f", &xpoints[i], &ypoints[i]);
          sigmax=sigmax+xpoints[i];
          sigmay=sigmay+ypoints[i];
          sigmaxy=sigmaxy+xpoints[i]*ypoints[i];
          sigmaxsquared=sigmaxsquared+(float)pow(xpoints[i],2);

     }
```

```
        /* Print out the points entered */
        printf("points are \n");
        for(i=0;i<points;i++)
        {
                printf(" \n");
                printf("%f %f", xpoints[i], ypoints[i]);
                fprintf(fp,"%lf\t%lf\n",xpoints[i], ypoints[i]);

        }
        printf(" \n");
        fltcnt=(float)points; /* Copy to fltcnt as a float type */

        /* Calculation of (xbar,ybar) - the mean points*/
        /* and sxy and sxx from the formulas*/
        xbar=sigmax/fltcnt;
        ybar=sigmay/fltcnt;
        sxy=(1/fltcnt)*sigmaxy-xbar*ybar;
        sxx=(1/fltcnt)*sigmaxsquared-xbar*xbar;

        /* Calculation of b and a from the formulas */
        b=sxy/sxx;
        a=ybar-b*xbar;

        /* Print the equation of the regression line */

        printf("Equation of regression line y on x  is\n ");
        printf(" y=%f + %fx", a,b);
        printf(" \n");

    }
    fclose(fp);
}
```

The following data for x values and y values was entered to the program and produced the following graph.

Data for car depreciation (8 points)

Age	Value ($1000)
2.5	11.5
3.0	10.6
3.5	9.2
4.0	7.8
4.5	6.1
5.0	4.7
5.5	3.9
6.0	1.8

The line of regression calculated by the program was

$$y = 18.7488 - 2.776x$$

This is plotted as the black line passing through the points.

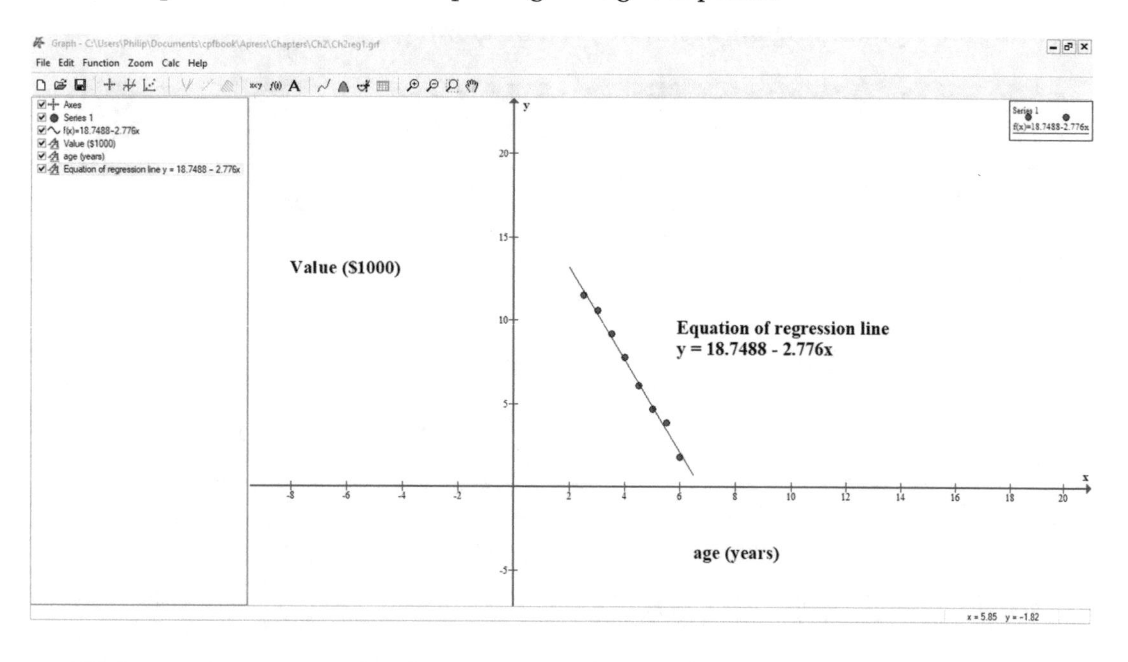

We have seen an example of negative correlation where the slope of the graph goes down from the left. The next example shows positive correlation. This shows the number of items produced by a company and it relates then to the company's costs.

No. of items (n)	Production costs (p) in $100
21	52
39	75.4
48	87.1
24	58.5
72	115.7
75	124.8
15	48.1
35	68.9
62	107.9
81	132.6
12	45.5
56	97.5

Using the regyonx program again, we get the following:

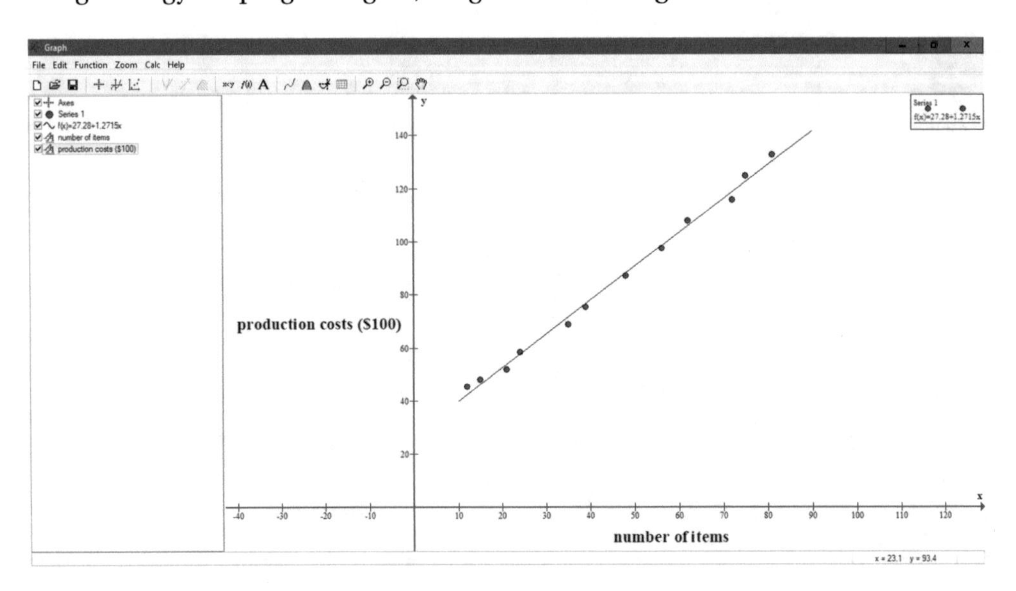

The regression line for this data was

$$y = 27.281 + 1.2715x$$

2.1 Capital Asset Pricing Model

This uses regression to model the relationship between expected returns and market risk. We can then use it to predict an expected return from the market risk.

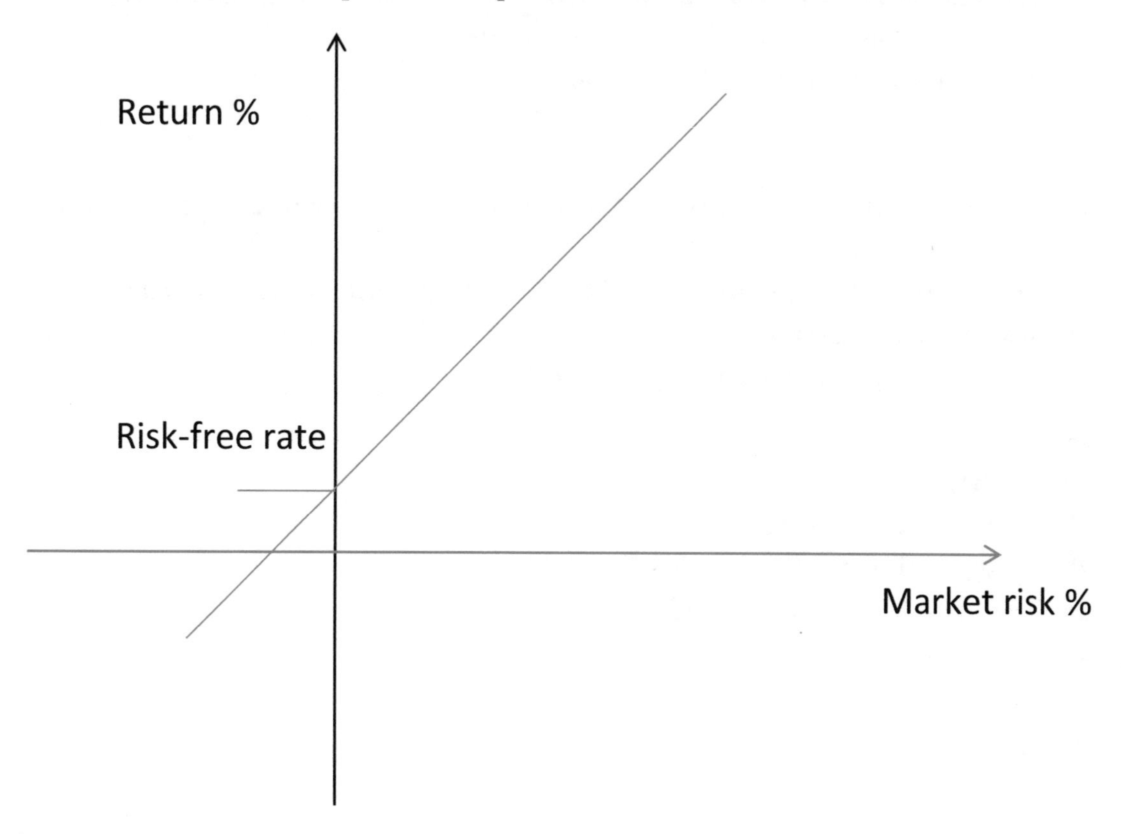

2.2 CAPM Illustration

The formula used for Capital Asset Pricing Model (CAPM) is

Expected return = Risk-free rate + (Beta * Market risk premium)

The risk-free rate is, for example, government securities which are generally thought to have zero risk.

Market risk premium (Rm – Rf) is the return you expect to get in the future.

Beta is a measure of the volatility of the investment compared to the market. If the beta of a stock is 1.5, then it would give a 150% change.

So CAPM takes into account the market risk rather than just the specific risk of your investment. It measures past performance to measure beta.

We can use our y on x regression formula

$$y = a + bx$$

which corresponds to our CAPM formula earlier where b in this formula corresponds to the beta in our CAPM formula.

So we can use our regyonx program for our CAPM calculation with just slight modification in the user requests.

The code for this program is shown as follows:

```
/* capm.c */
/*    CAPM  */
/*    User enters points.*/
/*    Regression of y on x calculated */
#define _CRT_SECURE_NO_WARNINGS
#include <stdio.h>
#include <math.h>
main()
{
     FILE *fp;

     float xpoints[12],ypoints[12];
     float sigmax,sigmay,sigmaxy,sigmaxsquared,xbar,ybar;
     float fltcnt,sxy,sxx,b,a;
     int i,points;
     fp=fopen("capm.dat","w");

     printf("enter number of points (max 12 ) \n");
     scanf("%d", &points);
     if(points>12)
     {
          printf("error - max of 12 points\n");

     }
```

```
else
{
    sigmax=0.0;
    sigmay=0.0;
    sigmaxy=0.0;
    sigmaxsquared=0.0;

    /* User enters points from scatter graph */
    for(i=0;i<points;i++)
    {
        printf("enter point (market percent change and share percent
        change separated by space) \n");
        scanf("%f %f", &xpoints[i], &ypoints[i]);
        sigmax=sigmax+xpoints[i];
        sigmay=sigmay+ypoints[i];
        sigmaxy=sigmaxy+xpoints[i]*ypoints[i];
        sigmaxsquared=sigmaxsquared+(float)pow(xpoints[i],2);

    }
    printf("points are \n");
    for(i=0;i<points;i++)
    {
        printf(" \n");
        printf("%f %f", xpoints[i], ypoints[i]);
        fprintf(fp,"%lf\t%lf\n",xpoints[i], ypoints[i]);
    }
    printf(" \n");
    fltcnt=(float)points;

    /* Calculation of (xbar,ybar) - the mean points*/
    /* and sxy and sxx from the formulas*/
    xbar=sigmax/fltcnt;
    ybar=sigmay/fltcnt;
    sxy=(1/fltcnt)*sigmaxy-xbar*ybar;
    sxx=(1/fltcnt)*sigmaxsquared-xbar*xbar;
```

```
        /* Calculation of b and a from the formulas */
        b=sxy/sxx;
        a=ybar-b*xbar;

        /* Print the equation of the regression line */

        printf("Equation of regression line y on x  is\n ");
        printf(" y=%f + %fx", a,b);
        printf(" \n");

        printf(" beta is %f", b);
    }
fclose(fp);
}
```

If we input the following historical data,

Market risk	Risk-free rate
1.500000	4.500000
2.000000	3.900000
2.100000	5.100000
1.900000	3.600000
-0.600000	-1.000000
-1.200000	-1.300000
-2.000000	-2.400000

we get the following graph:

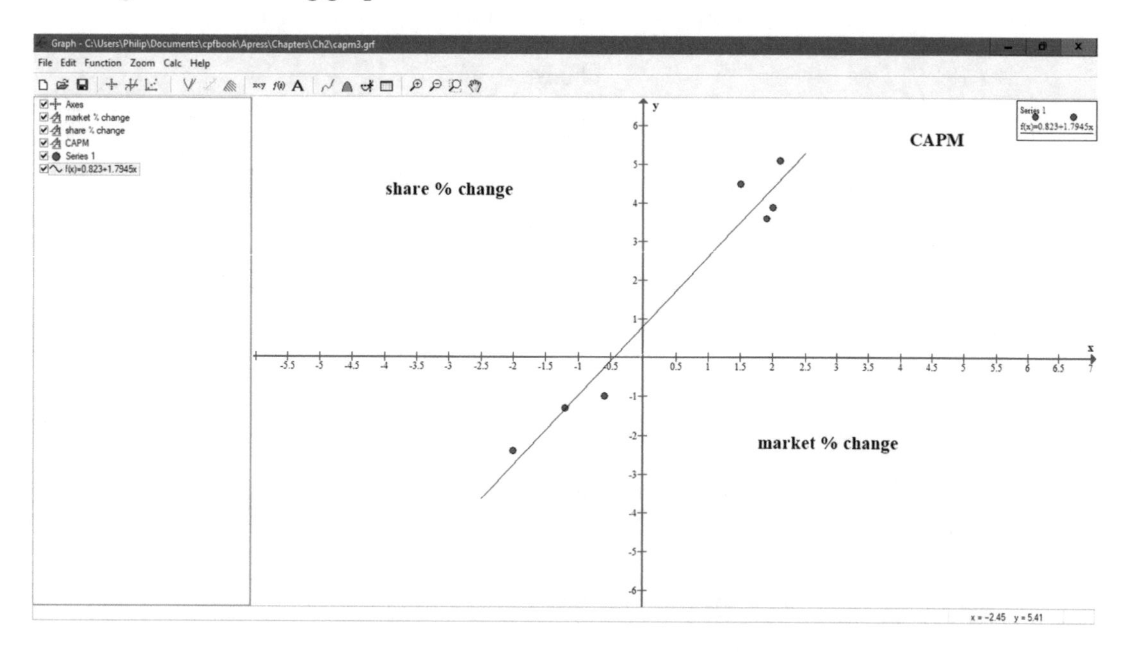

Here, the program calculates the regression line to be

$$y=0.822918 + 1.794479x$$

and the gradient of the line, which is beta, to be 1.794479.

EXERCISES

1. Use the data from the car value problem to calculate x on y. You can modify the program of y on x for this. Rename the output file "regxony". Print the regression line and the x and y points.
2. A company collates the number of units it produces with the cost of producing the units. The data is given in the following.

Units	Cost ($1000)
15	59.5
30	85
55	124.1
75	158.1
10	52.7
25	71.4
45	110.5
70	146.2
20	64.6
35	91.8
60	137.7
80	163.2

Use the regyonx program to produce a graph of the points and the regression line. Use your regression line to find the estimated cost of 50 units.

3. A chemical company varies the amount of a chemical agent injected into the blood and measures the resulting cholesterol level. Use the regyonx program with the x values as the agent level and the y values as the cholesterol level to produce a graph of points and the regression line. What conclusion can be drawn from this?

CHAPTER 3

PMCC

3.1 Theory

In Chapter 2 we saw positive and negative correlation and we saw the correlation to varying degrees. We could say that the correlation was strong if our points for our graph were close to the regression line.

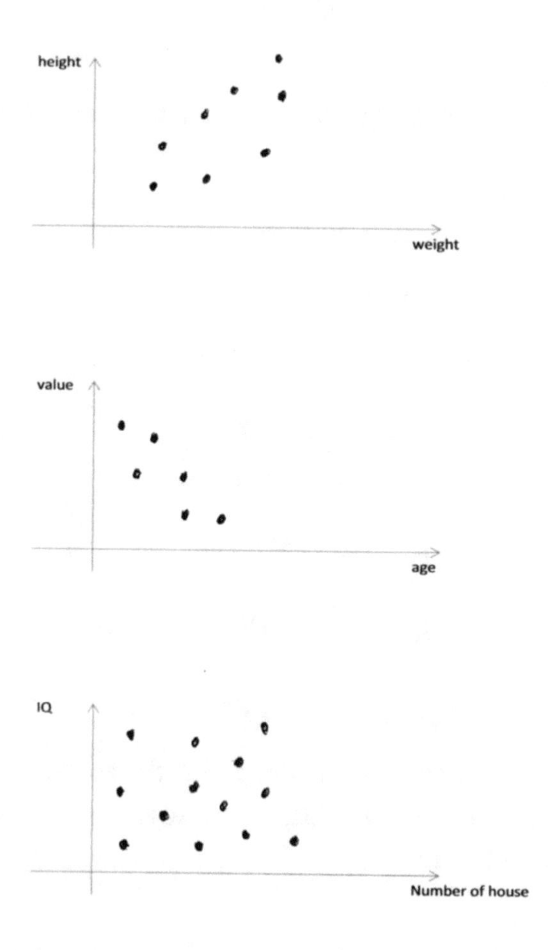

Figure 3-1. *Levels of correlation*

P. Joyce, *Practical Numerical C Programming*, https://doi.org/10.1007/978-1-4842-6128-6_3

In the preceding diagrams in Figure 3-1, you can see the two positive and negative correlation examples we used, but the third diagram is a bit strange. We are trying to relate a person's IQ to their house or apartment number. As you would think, there is no relationship between these and this is shown by the fact that all of the points are scattered all over the graph.

Rather than just saying that the correlation is strong or weak, we can assign a number to it. If, say, the points for our positive correlation are actually on the line of regression, then this is perfect correlation. We define a number for this. It is called the product moment correlation coefficient (PMCC), and for our case of perfect positive correlation, the PMCC is +1. If we had perfect negative correlation, then the PMCC would be -1. For the case earlier, with no correlation, the PMCC is 0.

The formula for the PMCC, **r**, is

$$r = S_{xy} / (S_x * S_y)$$

where $S_x = \sqrt{S_{xx}}$ and $S_y = \sqrt{S_{yy}}$

The values of S_{xx} and S_{yy} are just the same as we used in our regression programs from the previous chapter.

$$S_{xx} = \sum(x - \bar{x})^2$$

$$S_{yy} = \sum(y - \bar{y})^2$$

and

$$S_{xy} = \sum xy - (\sum x \sum y) / n$$

The more usable formats for these three formulas are

$$S_{xx} = \sum x^2 - (\sum x)^2/n$$

$$S_{yy} = \sum y^2 - (\sum y)^2/n$$

$$S_{xy} = \sum xy - (\sum x \sum y) / n$$

The PMCC is important as we can use it to determine the accuracy of our estimates of values from our regression lines. If r, the PMCC, is close to 1 or -1, then the values we read from the graphs will be accurate.

3.2 Manual Calculation of PMCC

We are going to use the six formulas from our theory of PMCC to find the value for the Car Depreciation problem. The six formulas are labeled as follows:

$$r = S_{xy} /(S_x * S_y) \quad (1)$$

$$\text{where } S_x = \sqrt{S_{xx}} \quad (2)$$

$$\text{and } S_y = \sqrt{S_{yy}} \quad (3)$$

$$S_{xx} = \sum x^2 - (\sum x)^2/n \quad (4)$$

$$S_{yy} = \sum y^2 - (\sum y)^2/n \quad (5)$$

$$S_{xy} = \sum xy - (\sum x \sum y) /n \quad (6)$$

We can now use the preceding formulas 1, 2, 3, 4, 5, and 6 to calculate the product moment correlation coefficient (r) for the Car Depreciation graph from Chapter 2. The x and y points from that graph are as follows:

Data for car depreciation (8 points)

x	y
2.5	11.5
3.0	10.6
3.5	9.2
4.0	7.8
4.5	6.1
5.0	4.7
5.5	3.9
6.0	1.8

Using these values in our six formulas, we get

$$\sum \mathbf{x} = 2.5 + 3.0 + 3.5 + 4.0 + 4.5 + 5.0 + 5.5 + 6.0 = 34$$

$$\sum \mathbf{y} = 11.5 + 10.6 + 9.2 + 7.8 + 6.1 + 4.7 + 3.9 + 1.8 = 55.6$$

$$\sum \mathbf{xy} = 2.5*11.5 + 3.0*10.6 + 3.5*9.2 + 4.0*7.8 + 4.5*6.1 + 5.0*4.7 + 5.5*3.9 + 6.0*1.8$$

$$= 28,75 + 31.8 + 32.2 + 31.2 + 27.45 + 23.5 + 21.45 + 10.8$$

$$= 207.15$$

$$\sum x^2 = 2.5^2 + 3.0^2 + 3.5^2 + 4.0^2 + 4.5^2 + 5.0^2 + 5.5^2 + 6.0^2$$

$$= 6.25 + 9 + 12.25 + 16 + 20.25 + 25 + 30.25 + 36$$

$$= 155$$

$$\sum y^2 = 11.5^2 + 10.6^2 + 9.2^2 + 7.8^2 + 6.1^2 + 4.7^2 + 3.9^2 + 1.8^2$$

$$= 132.25 + 112.36 + 84.64 + 60.84 + 37.21 + 22.09 + 15.21 + 3.24$$

$$= 467.84$$

From our values of $\sum x$ and $\sum y$, we get

$$\bar{x} = \sum x / 8 = 34/8 = 4.25$$

$$\bar{y} = \sum y / 8 = 55.6/8 = 6.95$$

From our values of $\sum x^2$, $\sum y^2$, and $\sum xy$, we get

$$S_{xx} = \sum x^2 - (\sum x)^2/n$$

$$= 155 - 34^2 / 8 = 10.5$$

$$S_{yy} = \sum y^2 - (\sum y)^2/n$$

$$= 467.84 - 55.6^2 / 8 = 81.42$$

$$S_{xy} = \sum xy - (\sum x \sum y) / n$$

$$= 207.15 - 34*55.6 / 8 = -29.15$$

So we can now write

$$S_x = \sqrt{S_{xx}} = 3.24037$$

$$S_y = \sqrt{S_{yy}} = 9.0233$$

Using these values for PMCC,

$$r = S_{xy} /(S_x * S_y)$$

$$= -29.15/(3.24037*9.0233)$$

$$= -0.996961$$

So our value for the product moment correlation coefficient for the Car Depreciation problem is -0.996961. This is very close to -1 which would be perfect negative correlation.

In some textbooks a slightly different notation for our six equations is used. The two notations are called the "small S format" and the "big S format." The equations earlier are in the "big S" format. In the "small S" format, equations 4, 5, and 6 become

$$sxx = (\sum x^2)/n - (\bar{x})^2 \quad (4)'$$

$$syy = (\sum y^2)/n - (\bar{y})^2 \quad (5)'$$

$$sxy = (\sum xy)/n - \bar{x} * \bar{y} \quad (6)'$$

The relationship between the "big S" and "small S" is

$$S_{xx} = n^* sxx$$

$$S_{yy} = n^* syy$$

$$S_{xy} = n^* sxy$$

The rest of the equations are the same. As we are using these equations to find the PMCC, r, and the regression equation variables b and d, then the n term that differentiates between the two formats cancels by the division used in the formulas.

So the definitions for the "small S" formulas are as the "big S" formulas

$$sx = \sqrt{sxx}$$

$$sy = \sqrt{syy}$$

$$r = sxy/(sx*sy)$$

$$b = sxy/sxx$$

$$d = sxy/syy$$

3.3 PMCC Program

We can just amend our regression programs to calculate the value of the PMCC using the preceding formulas.

We can then run the program using the data for our car depreciation example in our regression chapter. The data for this is as follows:

Data for car depreciation (8 points)

2.5	11.5
3.0	10.6
3.5	9.2
4.0	7.8
4.5	6.1
5.0	4.7
5.5	3.9
6.0	1.8

And an example of the code we can use for PMCC is shown in the following:

```
/*Product moment correlation coefficient */
#define _CRT_SECURE_NO_WARNINGS
#include <stdio.h>
#include <math.h>
main()
{
      double xpoints[10], ypoints[10];

/* Variables are named as in the formulas used in the text of the chapter
*/
/* e.g., Σx is called "sigmax" */

      double sigmax, sigmay, sigmaxsquared, sigmaysquared, xbar, ybar,
      sigmaxy;
      double sxy, sxx, syy, sx, sy, r;
      int i, points; /* User-entered number of scatter graph points */
      double fltcnt; /* Number of points as a double variable */
      FILE *fp;
      fp=fopen("pmccf.dat","w");
      /* User enters number of points in scatter graph */
      printf("enter number of points (max 10 ) \n");
      scanf("%d", &points);
      if (points > 10)
      {
```

```
        printf("error - max of 10 points\n");
    }
    else
    {
    /* Variables used for summing data values cleared to zero */
        sigmax = 0.0;
        sigmay = 0.0;
        sigmaxy = 0.0;
        sigmaxsquared = 0.0;
        sigmaysquared = 0.0;

        /* User enters points in scatter graph */
        for (i = 0;i < points;i++)
        {
            printf("enter point (x and y separated by space) \n");
            scanf("%lf %lf", &xpoints[i], &ypoints[i]);
            /* Totals incremented by x and y points */
            sigmax = sigmax + xpoints[i];
            sigmay = sigmay + ypoints[i];
            sigmaxy = sigmaxy + xpoints[i] * ypoints[i];
            sigmaxsquared = sigmaxsquared + pow(xpoints[i], 2);
            sigmaysquared = sigmaysquared + pow(ypoints[i], 2);
        }
        printf("points are \n");
        for (i = 0;i < points;i++)
        {
            printf(" \n");
            printf("%lf %lf", xpoints[i], ypoints[i]);
            fprintf(fp,"%lf\t%lf\n",xpoints[i], ypoints[i]);
        }
        printf(" \n");

        /* Convert number of points as a double variable */
        /* for use in the formulas. Store this in variable fltcnt */

        fltcnt = (double)points;
```

```
        /* variables in PMCC formula calculated */
        xbar = sigmax / fltcnt;
        ybar = sigmay / fltcnt;

        syy = (1 / fltcnt)*sigmaysquared - ybar * ybar;

        sxx = (1 / fltcnt)*sigmaxsquared - xbar * xbar;
        sx = sqrt(sxx);
        sy = sqrt(syy);
        sxy = (1 / fltcnt)*sigmaxy - xbar * ybar;

        /* PMCC value calculated */
        r = sxy / (sx*sy);
        printf("r is %lf", r);
    }

    fclose(fp);
}
```

The resulting graph of points and the calculated PMCC are shown in Figure 3-2 as follows. The points given earlier are entered into the Graph package manually using "Function" ➤ "Insert point series" within the Graph package command line. The value for PMCC calculated by the program is the same as our manual calculation in this chapter.

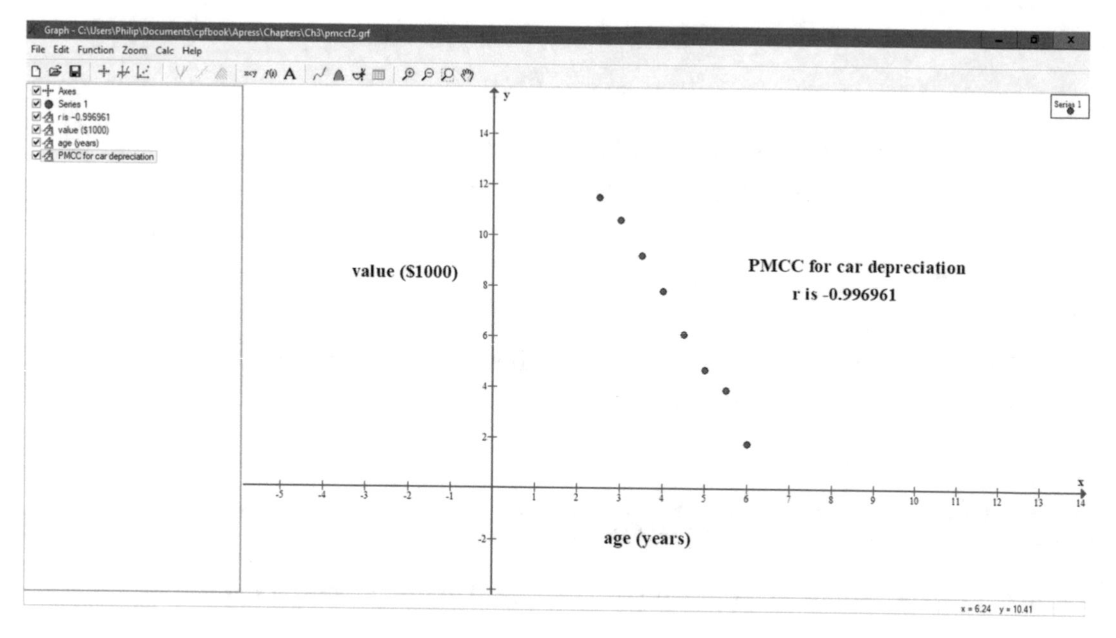

Figure 3-2. *PMCC of Car Depreciation*

We see that the PMCC is -0.996961 which is very close to -1 so showing very strong negative correlation.

3.4 Comparison of the Two Regression Lines

From our two equations of regression of y on x used in the previous chapter,

$$y = a+bx$$

where $b = S_{xy}/ S_{xx}$ and $a = \bar{y} - b\bar{x}$
and of x on y

$$x = c + dy$$

where $d = S_{xy}/ S_{yy}$ and $c = \bar{x} - d\bar{y}$
we can see from the preceding definitions of **b** and **d** that

$$\mathbf{b*d} = (S_{xy}/ S_{xx})*(S_{xy}/S_{yy})$$

$$= (S_{xy}/ S_x S_x)*(S_{xy}/ S_y S_y)$$

(as $S_{xx} = S_x S_x$ and $S_{yy} = S_y S_y$ from the preceding formulas)

$$\text{so } b*d =(S_{xy} * S_{xy})/(S_x S_x * S_y S_y)$$

$$\text{so } b*d = (S_{xy}/(S_x S_y))^2$$

$$\text{but } \mathbf{r^2} = (Sxy/(\,Sx\,Sy))^2$$

$$\text{so } r = \sqrt{(bd)}$$

The regression line of y on x y = a + bx has gradient b.

The regression line of x on y x = c + dy has gradient 1/d.

If $r^2 = 1$

as $r^2 = bd$

then bd = 1 so b = 1/d.

So both regression lines have the same gradient, and as they both pass through $\overline{x}$ and $\overline{y}$, then the two lines must be identical.

3.5 Manual Calculation of the Two Regression Lines

From our two equations of regression of y on x used in the previous chapter,

$$y = a+bx$$

where b = sxy/ sxx and a = $\overline{y}$ - b$\overline{x}$
and of x on y

$$x = c + dy$$

where d = sxy/ syy and c = $\overline{x}$ - d $\overline{y}$
and

$$sxx = (\sum x^2)/n - (\,\overline{x}\,)^2 \qquad (4)'$$

$$syy = (\sum y^2)\,/n - (\,\overline{y}\,)^2 \qquad (5)'$$

$$sxy = (\sum xy)\,/n - \overline{x} * \overline{y} \qquad (6)'$$

for the points

x	y
0.000000	1.000000
1.000000	3.000000
3.000000	7.000000
4.000000	9.000000
6.000000	13.000000

we get

$$\sum \mathbf{x} = 0+1+3+4+6 = 14$$

$$\sum \mathbf{y} = 1+3+7+9+13 = 33$$

$$\sum \mathbf{xy} = 0*1 + 1*3 + 3*7 + 4*9 + 6*13 = 138$$

$$\sum \mathbf{x}^2 = 0+1+9+16+36 = 62$$

$$\sum \mathbf{y}^2 = 1+9+49+81+169 = 309$$

From our values of $\sum x$ and $\sum y$, we can write

$$\bar{x} = \sum x / 5 = 14/5 = 2.8$$

$$\bar{y} = \sum y / 5 = 33/5 = 6.6$$

From our values of $\sum x$, $\sum y$, and $\sum xy$, we can write

$$\mathbf{sxx} = (\sum x^2)/n - (\bar{x})^2 = 0.2*62 - 2.8^2 = 4.56 \quad (4)'$$

$$\mathbf{syy} = (\sum y^2)/n - (\bar{y})^2 = 0.2*309 - 6.6^2 = 18.24 \quad (5)'$$

$$\mathbf{sxy} = (\sum xy)/n - \bar{x}*\bar{y} = 0.2*138 - 2.8*6.6 = 9.12 \quad (6)'$$

So we can now write

$$\mathbf{sx} = \sqrt{sxx} = \sqrt{4.56} = 2.13541565$$

$$\mathbf{sy} = \sqrt{syy} = \sqrt{18.24} = 4.270831301$$

$$\mathbf{sxy} = 9.12$$

So $\mathbf{r} = sxy/(sx*sy)$

$= 9.12/(2.13541565*4.270831301)$

$= 1$ (So PMCC = 1)

Also from the two regression equations

$$y = a + bx \text{ and } x = c + dy$$

$$b = sxy/sxx = 9.12/4.56 = 2$$

$$a = \bar{y} - b\bar{x} = 6.6 - 2*2.8 = 1$$

$$d = sxy/syy = 9.12/18.24 = 0.5$$

$$c = \bar{x} - d\bar{y} = 2.8 - 0.5*6.6 = -0.5$$

So the two regression equations

$$y = a + bx \text{ and } x = c + dy$$

in this case are $y = 1 + 2x$ and $x = -0.5 + 0.5y$.

If we rearrange $x = -0.5 + 0.5y$, we get $0.5y = x + 0.5$ or $y = 2x + 1$.

So the y on x and x on y regression equations are identical.

3.6 Program for the Two Regression Lines

The following code creates both regression lines, y on x and x on y. Run the following program and input the following points into it.

x	y
0.000000	1.000000
1.000000	3.000000
3.000000	7.000000
4.000000	9.000000
6.000000	13.000000

The points produce the line $y = 2x + 1$.

```
/* regbothip.c */
/*      regression */
/*      User enters points.*/
/*      Regression of y on x and x on y calculated */
#define _CRT_SECURE_NO_WARNINGS
#include <stdio.h>
#include <math.h>
main()
{
FILE *fp;

       double xpoints[16];
       double ypoints[16];
/* Variables are named as in the formulas used in the text of the chapter */
```

```
/* e.g., Σx is called "sigmax" */

        double sigmax,sigmay,sigmaxy,sigmaxsquared,sigmaysquared,xbar,ybar;
        double fltcnt,sxy,sxx,syy,b,a,c,d, sx, sy, r;
        int i,points;
        fp=fopen("regbothip.dat","w");

        /* User enters points in scatter graph */

        printf("enter number of points (max 16 ) \n");
        scanf("%d", &points);
        if(points>16)
        {
                printf("error - max of 16 points\n");

        }
        else
        {
                sigmax=0.0;
                sigmay=0.0;
                sigmaxy=0.0;
                sigmaxsquared=0.0;
                sigmaysquared=0.0;

                /* User enters points from scatter graph */
                for(i=0;i<points;i++)
                {
                        printf("enter point (x and y separated by space) \n");
                        scanf("%lf %lf", &xpoints[i], &ypoints[i]);
                        sigmax=sigmax+xpoints[i];
                        sigmay=sigmay+ypoints[i];
                        /* Totals incremented by x and y points */
                        sigmaxy=sigmaxy+xpoints[i]*ypoints[i];
                        sigmaxsquared=sigmaxsquared+pow(xpoints[i],2);
                        sigmaysquared=sigmaysquared+pow(ypoints[i],2);
                }
                printf("points are \n");
                for(i=0;i<points;i++)
```

```
{
        printf(" \n");
        printf("%lf %lf", xpoints[i], ypoints[i]);
        fprintf(fp,"%lf\t%lf\n",xpoints[i], ypoints[i]);
}
printf(" \n");

/* Convert number of points as a double variable */
/* for use in the formulas. Store this in variable fltcnt */

fltcnt=(double)points;

/* Calculation of (xbar,ybar) - the mean points*/
/* and sxy and sxx from the formulas*/
xbar=sigmax/fltcnt;
ybar=sigmay/fltcnt;
sxy=(1/fltcnt)*sigmaxy-xbar*ybar;
sxx=(1/fltcnt)*sigmaxsquared-xbar*xbar;
syy=(1/fltcnt)*sigmaysquared-ybar*ybar;

sx = sqrt(sxx);
sy = sqrt(syy);

/* calculation of b and a from the formulas */
/* described earlier in this chapter */
b=sxy/sxx;
a=ybar-b*xbar;

/* Print the equation of the regression line */

printf("Equation of regression line y on x  is\n ");
printf(" y=%lf + %lfx", a,b);
printf(" \n");

/* Calculation of d and c from the formulas */
d=sxy/syy;
c=xbar-d*ybar;

/* Regression line */
printf("Equation of regression line x on y  is\n ");
```

```
                printf(" x=%lf + %lfy", c,d);
        }
        /* PMCC value calculated */
        r = sxy / (sx*sy);
        printf("\nr is %lf", r);
        fclose(fp);
}
```

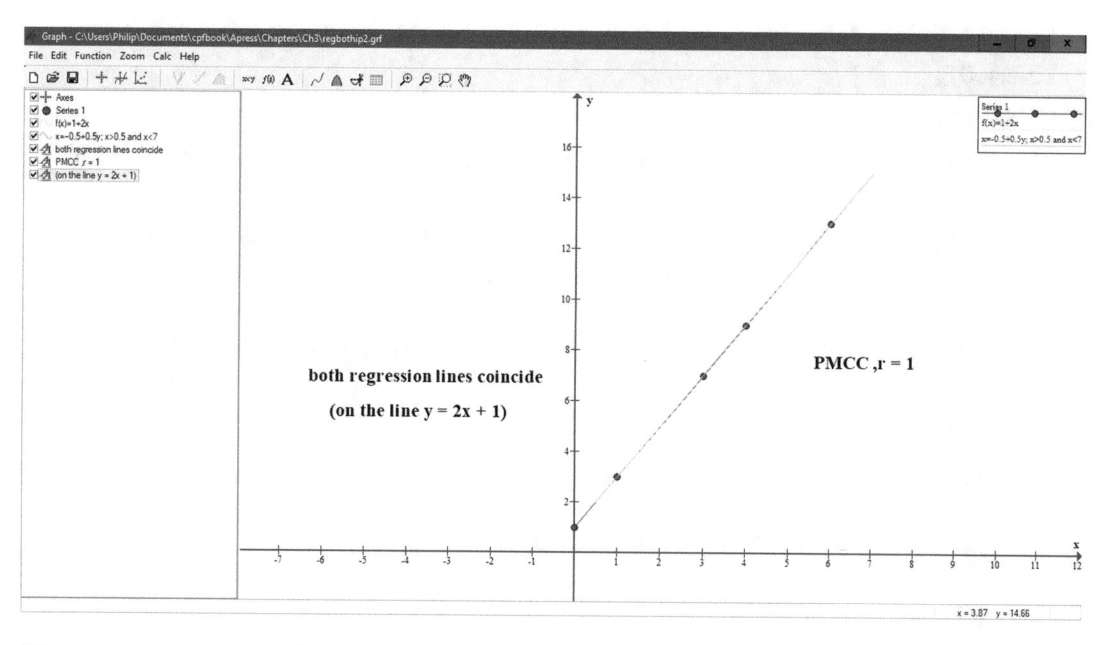

Figure 3-3. *PMCC of +1*

The preceding graph in Figure 3-3 shows the output of the program. This shows perfect positive correlation (r = 1). Also it shows the two regression lines overlap.

We can demonstrate a high PMCC and a low PMCC along with the corresponding y on x and x on y regression lines.

Here is the code. The data is preset so that you don't have to enter the points.

```
/* regboth.c */
/*       Regression */
/*       Points are preset into xpoints and ypoints */
/*       Regression of y on x and x on y calculated */
```

```
#define _CRT_SECURE_NO_WARNINGS
#include <stdio.h>
#include <math.h>
main()
{
       FILE *fp;
       /* Preset points */

double xpoin
ts[16]={1.0,3.0,4.0,5.0,7.0,17.0,9.0,10.0,12.0,14.0,16.0,17.0,19.0,7.0,
12.0,14.0};
double ypoin
ts[16]={8.0,11.0,14.0,7.0,2.0,19.0,13.0,3.0,10.0,5.0,12.0,19.0,15.0,16.0,
16.0,8.0};
/* Variables are named as in the formulas used in the text of the chapter */
/* e.g., Σx is called "sigmax" */

       double sigmax,sigmay,sigmaxy,sigmaxsquared,sigmaysquared,xbar,ybar;
       double fltcnt,sxy,sxx,syy,b,a,c,d, sx, sy, r;
       int i,points;
       fp=fopen("regboth2a.dat","w");

       points = 16; /* Number of points is 16 */


       sigmax=0.0;
       sigmay=0.0;
       sigmaxy=0.0;
       sigmaxsquared=0.0;
       sigmaysquared=0.0;

       /* Points are preset in xpoints and ypoints */
       for(i=0;i<points;i++)
       {
              sigmax=sigmax+xpoints[i];
              sigmay=sigmay+ypoints[i];
              sigmaxy=sigmaxy+xpoints[i]*ypoints[i];
```

```
        sigmaxsquared=sigmaxsquared+pow(xpoints[i],2);
        sigmaysquared=sigmaysquared+pow(ypoints[i],2);
}
printf("points are \n");
for(i=0;i<points;i++)
{
        printf(" \n");
        printf("%lf %lf", xpoints[i], ypoints[i]);
        fprintf(fp,"%lf\t%lf\n",xpoints[i], ypoints[i]);
}
printf(" \n");
fltcnt=(double)points;

/* Calculation of (xbar,ybar) - the mean points*/
/* and sxy and sxx from the formulas*/
xbar=sigmax/fltcnt;
ybar=sigmay/fltcnt;
sxy=(1/fltcnt)*sigmaxy-xbar*ybar;
sxx=(1/fltcnt)*sigmaxsquared-xbar*xbar;
syy=(1/fltcnt)*sigmaysquared-ybar*ybar;

sx = sqrt(sxx);
sy = sqrt(syy);

/* Calculation of b and a from the formulas */
b=sxy/sxx;
a=ybar-b*xbar;

/* Print the equation of the regression line */

printf("Equation of regression line y on x  is\n ");
printf(" y=%lf + %lfx", a,b);
printf(" \n");

/* Calculation of d and c from the formulas */
d=sxy/syy;
c=xbar-d*ybar;
```

```
        /* Regression line */
        printf("Equation of regression line x on y  is\n ");
        printf(" x=%lf + %lfy", c,d);

        /* PMCC value calculated */
        r = sxy / (sx*sy);
        printf("\nr is %lf", r);
        fclose(fp);
}
```

The points, two regression lines and PMCC value, are shown in Figure 3-4.

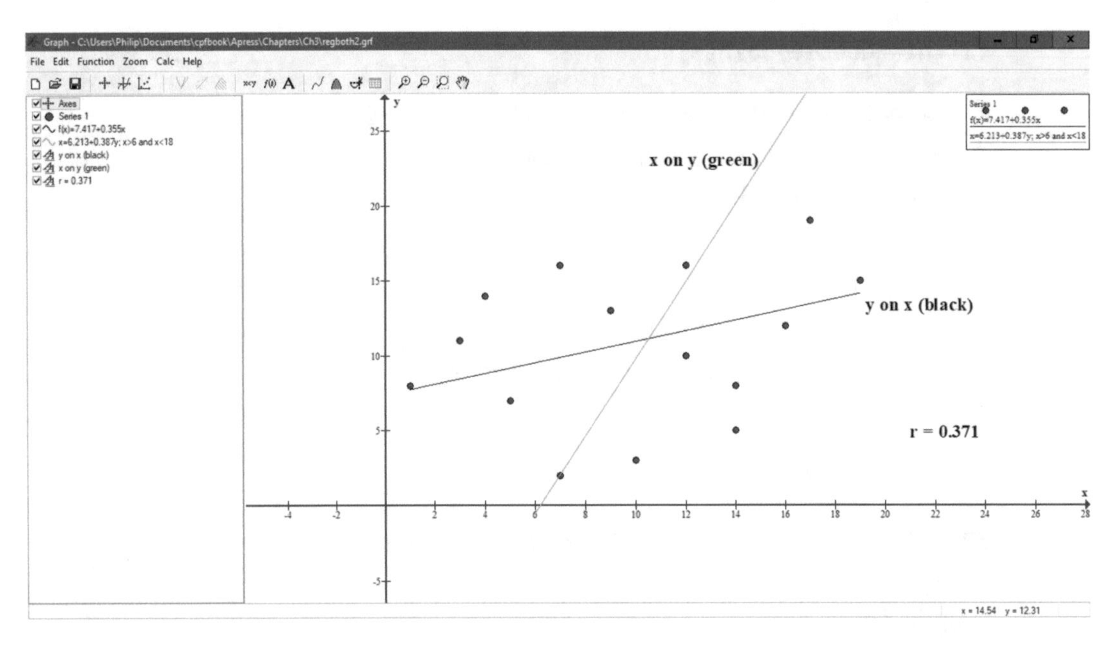

***Figure 3-4.** y on x and x on y comparison*

We can demonstrate points with a stronger correlation by using different points. We can use the same program as earlier but replace xpoints and ypoints with the following values:

```
xpoints[16] = {2.0,4.0,4.0,4.0,6.0,6.0,7.0,8.0,9.0,10.0,10.0,10.0,12.0,
13.0,14.0,15.0};

ypoints[16] = {4.0,2.0,5.0,8.0,4.0,10.0,6.0,12.0,14.0,8.0,11.0,17.0,19.0,
13.0,14.0,17.0};
```

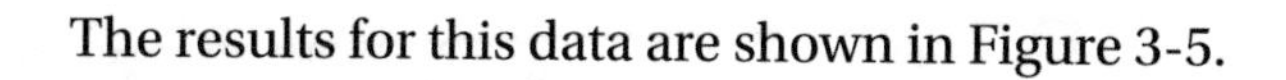

The results for this data are shown in Figure 3-5.

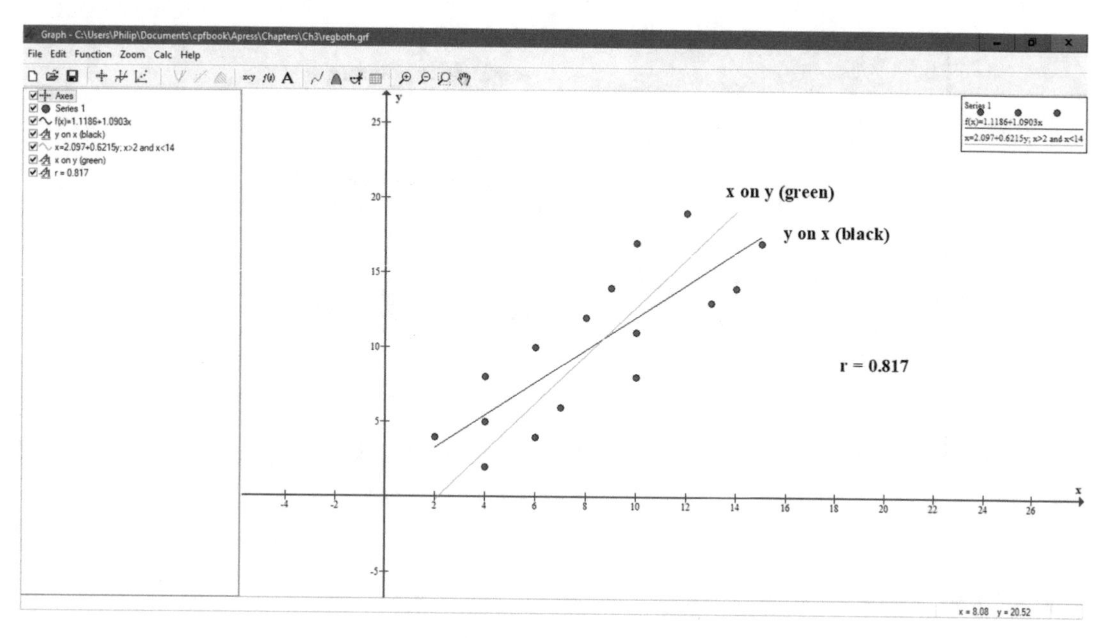

Figure 3-5. *y on x and x on y higher correlation*

You can see that the points are closer together, the two regression lines are closer together, and the PMCC value is 0.817 compared with 0.371 in the previous diagram.

Looking at the other extreme, we can demonstrate a PMCC value of zero.

In the following program, the x and y points are preset to

```
xpoints[16]={2.0,2.0,2.0,2.0,2.0,18.0,18.0,18.0,18.0,18.0,6.0,10.0,14.0,
6.0,10.0,14.0};
```

```
ypoints[16]={2.0,6.0,10.0,14.0,18.0,2.0,6.0,10.0,14.0,18.0,2.0,2.0,2.0,
18.0,18.0,18.0};
```

These points form a square.

Look at the following graph in Figure 3-6.

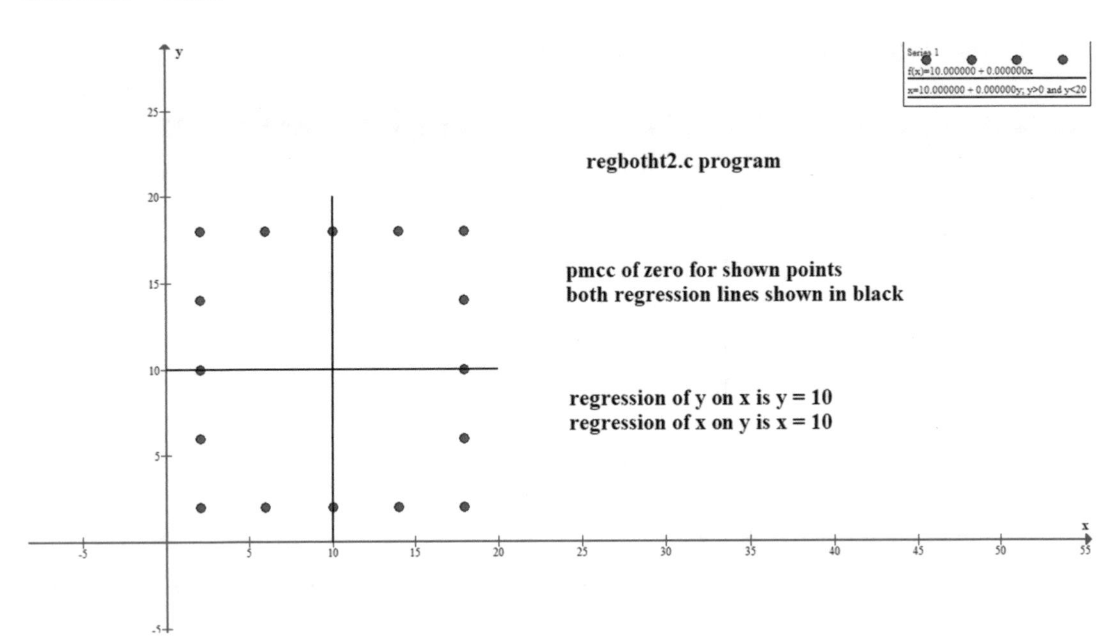

Figure 3-6. *PMCC = 0*

Here, the points form a square. The two regression lines are at right angles to each other and to the sides of the rectangle. The PMCC for this is zero. Looking at the points, you could say that there was correlation between them, but the PMCC value only looks for linear correlation of the whole set of points.

The program for this is as follows:

```
/* regbotht2.c */
/*        Regression */
/*        Preset points.*/
/*        Regression of y on x and x on y calculated */
/* Test for PMCC of zero */
/* Show appropriate regression lines */
#define _CRT_SECURE_NO_WARNINGS
#include <stdio.h>
#include <math.h>
main()
{
FILE *fp;
```

```
/* Preset points for x and y values */
float xpoints[16]={2.0,2.0,2.0,2.0,2.0,18.0,18.0,18.0,18.0,18.0,6.0,
10.0,14.0,6.0,10.0,14.0};
float ypoints[16]={2.0,6.0,10.0,14.0,18.0,2.0,6.0,10.0,14.0,18.0,
2.0,2.0,2.0,18.0,18.0,18.0};

float sigmax,sigmay,sigmaxy,sigmaxsquared,sigmaysquared,xbar,ybar;
float fltcnt,sxy,sxx,syy,b,a,c,d, sx, sy, r;
int i,points;

/* Open the output file */
fp=fopen("regbotht2.dat","w");

points = 16; /* Number of points fixed at 16 */

/* Preset storage variables to zero */
sigmax=0.0;
sigmay=0.0;
sigmaxy=0.0;
sigmaxsquared=0.0;
sigmaysquared=0.0;

/* Points from preset arrays */
/* Calculate sigmax,sigmay,sigmaxy,sigmaxsquared,sigmaysquared */
for(i=0;i<points;i++)
{

        sigmax=sigmax+xpoints[i];
        sigmay=sigmay+ypoints[i];
        sigmaxy=sigmaxy+xpoints[i]*ypoints[i];
        sigmaxsquared=sigmaxsquared+(float)pow(xpoints[i],2);
        sigmaysquared=sigmaysquared+pow(ypoints[i],2);
}

/* Print out the points and write them to the output file */
printf("points are \n");
```

```
for(i=0;i<points;i++)
{
        printf(" \n");
        printf("%f %f", xpoints[i], ypoints[i]);
        fprintf(fp,"% f\t%f\n",xpoints[i], ypoints[i]);
}
printf(" \n");
fltcnt=(float)points; /* Set float variable to count of points */

/* Calculation of (xbar,ybar) - the mean points*/
/* and sxy and sxx from the formulas*/
xbar=sigmax/fltcnt;
ybar=sigmay/fltcnt;
sxy=(1/fltcnt)*sigmaxy-xbar*ybar;
sxx=(1/fltcnt)*sigmaxsquared-xbar*xbar;
syy=(1/fltcnt)*sigmaysquared-ybar*ybar;

/* Calculate sx and sy from the formulas*/
sx = sqrt(sxx);
sy = sqrt(syy);

/* Calculation of b and a from the formulas */
b=sxy/sxx;
a=ybar-b*xbar;

/* Print the equations of the regression lines */

/* Regression line y on x */
printf("Equation of regression line y on x  is\n ");
printf(" y=%f + %fx", a,b);
printf(" \n");

/* Calculation of d and c from the formulas */
d=sxy/syy;
c=xbar-d*ybar;

/* Regression line x on y */
printf("Equation of regression line x on y  is\n ");
printf(" x=%f + %fy", c,d);
```

```
        /* PMCC value calculated */
        r = sxy / (sx*sy);
        printf("\nr is %f", r);
        fclose(fp);
}
```

EXERCISES

1. Use one of the PMCC programs to find PMCC data for the following points. You might want to change the name of the output file that the program uses so that you can keep it separate from previous files.

x	y
-4	16
-3	9
0	0
3	9
4	16

Find the value of PMCC for this. What do you notice about the shape of the pattern that the points make?

CHAPTER 4

Stock Price Prediction

4.1 Two Parts to Stock Price Changes

Stock price prediction combines the Brownian motion theory of physics with the Monte Carlo theory of statistical mathematics. Put simply, Brownian motion just models the random motion of particles in gases and liquids. If we assumed that there was a general drift upward and to the right and we plotted the particles' position in 2D over a period of seconds, the shape of the graph would be similar to that of the following graph in Figure 4-1 which shows stock price variation over a number of days.

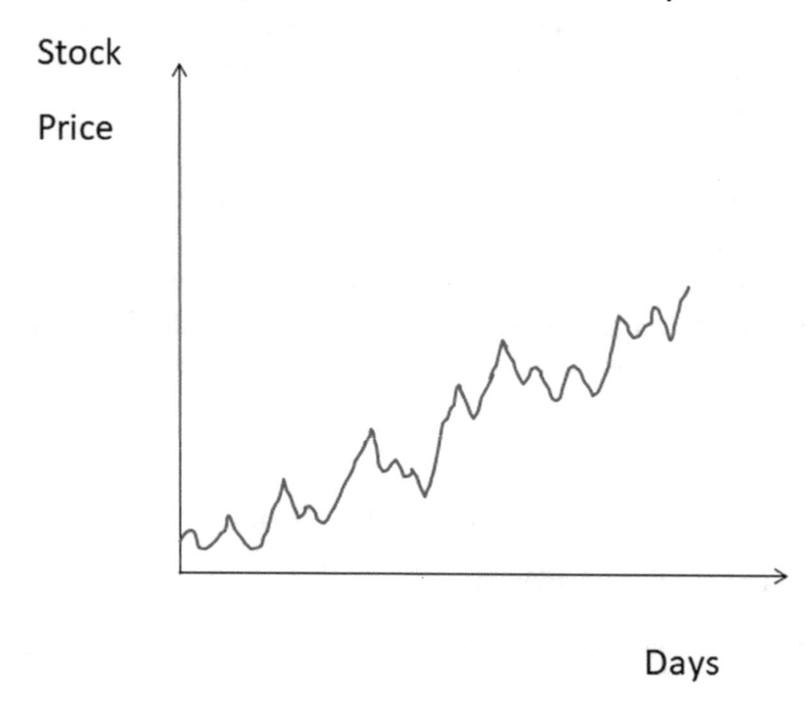

Figure 4-1. *Brownian Motion Illustration*

P. Joyce, *Practical Numerical C Programming*, https://doi.org/10.1007/978-1-4842-6128-6_4

This relates the general direction of movement (drift) and the random movement of particles/stock price.

In the stock prices case, we can say that the "drift" is the general change in the stock price over a period of time. Just glancing at the graph, we can see that this goes up. Although the graph goes up and down over a few days, the general trend of the graph is to go up. The short-term rise and fall just reflects the random daily volatility of the stock price.

So we can say that

Change in stock price = Drift + Random Change

The "Monte Carlo" part of the stock price prediction has applications in physics and mathematics. This is utilized in the Random Change part of the formula. It uses random numbers which can be generated by a computer program. This is the same kind of thing that you can get with a school calculator. Depending on the make of calculator you have, you would have a key, say RAN, which, when you press it, displays a random number between 0 and 1, say 0.883. The next time you press the key, you would get a different random number, say 0.393. If you switch your calculator off and on again and tried pressing RAN twice again, you would get two different numbers. The random number generator in a computer program works in a similar way, except that your decimal numbers will be to more decimal places.

The fact that the numbers generated are between 0 and 1 is not a problem. For instance, if you wanted your decimal numbers to be between 0 and 10, you just have to multiply the number your program generates by 10. So your number will be between 0 and 10. So in the preceding two cases, you would get 8.83 and 3.93. If you want numbers between 5 and 15, you can just add 5 to your 0 to 10 case, so for the two cases here, you would get 8.93 + 5 = 13.83 and 3.93 + 5 = 8.93. So you can generate random numbers within any range you want.

So we can write our previous equation

Change in stock price = Drift + Random Change

And we can rewrite this equation as

New Stock Price = Old Stock price *(A Factor involving Drift + Random Change)

So if we want to find today's predicted stock price in relation to yesterday's, then we can write

Today's Stock Price = Yesterday's Stock price * exp(Drift + Random Change)

where here the "Factor involving Drift + Random Change" involves the exponential function "exp" and becomes exp(Drift + Random Change).

4.2 Drift Part of Formula

We can find the drift using the formula

Drift = Average Daily Return - (Variance/2)

Variance is a measure of how much each value varies from the mean. The formula for this is shown in the following.

So if we have figures for the stock price for, say, the last 19 days, we can find the return for each day by relating yesterday's price to today's price and so on, giving us 18 daily return values. The formula for calculating this is

Periodic Daily Return = ln (Day's Price / Previous Day's Price)

The general formula for the variance

$$\text{Variance} = (\sum(x - \bar{x})^2) / n$$

where x is the value of each day's return and $\bar{x}$ is their average, and pdrx and pdr $\bar{x}$ are the PDR cases. The Greek letter $\sum$ is used to signify finding the sum of a set of values. In this case we take the mean value $\bar{x}$ from each x value in turn and square it and **n** is the number of daily return values. Then we sum all of these values.

In the PDR (Periodic Daily Return) case, we modify the formula to be

$$\text{PDRvariance} = (\sum(pdrx - pdr\bar{x})^2) / n$$

where x is the value of each day's return and $\bar{x}$ is their average, and PDRx and PDR $\bar{x}$ are the PDR cases. The Greek letter $\sum$ is used to signify finding the sum of a set of values. In this case we subtract the mean value $\bar{x}$ from each x value in turn and square it. The value of n is the number of daily return values which, as described earlier, will be 18. Then we sum all of these values.

4.3 Simple Example with 5 Day's Prices

To clarify this, let us take an example of stock price changes over 5 days.

Today's stock price ($)	= 22.8
Yesterday's price ($)	= 22.5
Previous day ($)	= 22.1
Day before ($)	= 22.9
First day ($)	= 21.7

Periodic Daily Return = ln (Day's Price / Previous Day's Price)

As we need the previous day's price for each periodic day return, we can only get four values here as we don't have the stock price for the day before the First day.

So applying the preceding formula for each day and its previous day, we get

PDR1 = ln (22.8/22.5) = 0.0129

PDR2 = ln (22.5/22.1) = 0.0178

PDR3 = ln (22.1/22.9) = -0.0356

PDR4 = ln (22.9/21.7) = 0.0535

The sum of these values is 0.0486.

So the average daily return is 0.0486 / 4 = 0.0121.

Now we need the formula to find the variance

PDRvariance = ($\sum$(pdrx - pdr$\overline{x}$)^2) / 4 (we have 4 PDR values in our case)

Applying this formula to our four values, we get

$(0.0129 - 0.0121)^2 + (0.0178 - 0.0121)^2 + (-0.0356 - 0.0121)^2 + (0.0535 - 0.0121)^2$

This is **($\sum$(pdrx - pdr$\overline{x}$)^2)** which is 0.00402238.

So **($\sum$(pdrx - pdr$\overline{x}$)^2) / 4** is 0.00402238 / 4 = 0.00105595 which is the **PDR variance.**

So from the formula

Drift = Average Daily Return - (Variance/2)

we get in our case

Drift = 0.0121 - 0.00105595/2 = 0.011572025

4.4 Random Change Part of Formula

Having found the drift, our next job is to find the Random Change component. For this we need the random number generated by the computer, but as explained earlier, we need to modify the value the computer gives us.

As in many cases, the variation about a mean value could be large or small but generally it will be somewhere in between. It is a bit like measuring the heights of men you pass in the street. Most men will be the average height with some variation, but occasionally you will get a very short man or a very tall man. We call this a “normal distribution” and the graph of this is a bell shape as shown in the following diagram. It is known that stock prices tend to follow a normal distribution.

The two straight lines mark the “standard deviation” points. Statistics say that 68% of the values lie between these two lines. The standard deviation, usually denoted by the Greek letter σ, is related to the variance as

$$\sigma = \sqrt{(\text{variance})}$$

For our stock market bell-shaped curve, our average value would be the most-likely value over a period of time.

The bell-shaped Normal Distribution curve is shown in Figure 4-2 with the two standard deviation lines denoted by “sd” at +1 and -1. The vertical axis line denoted the mean position, so 68% of the values lies between these two lines. This corresponds with our instinct that most adult people have a height close to the mean.

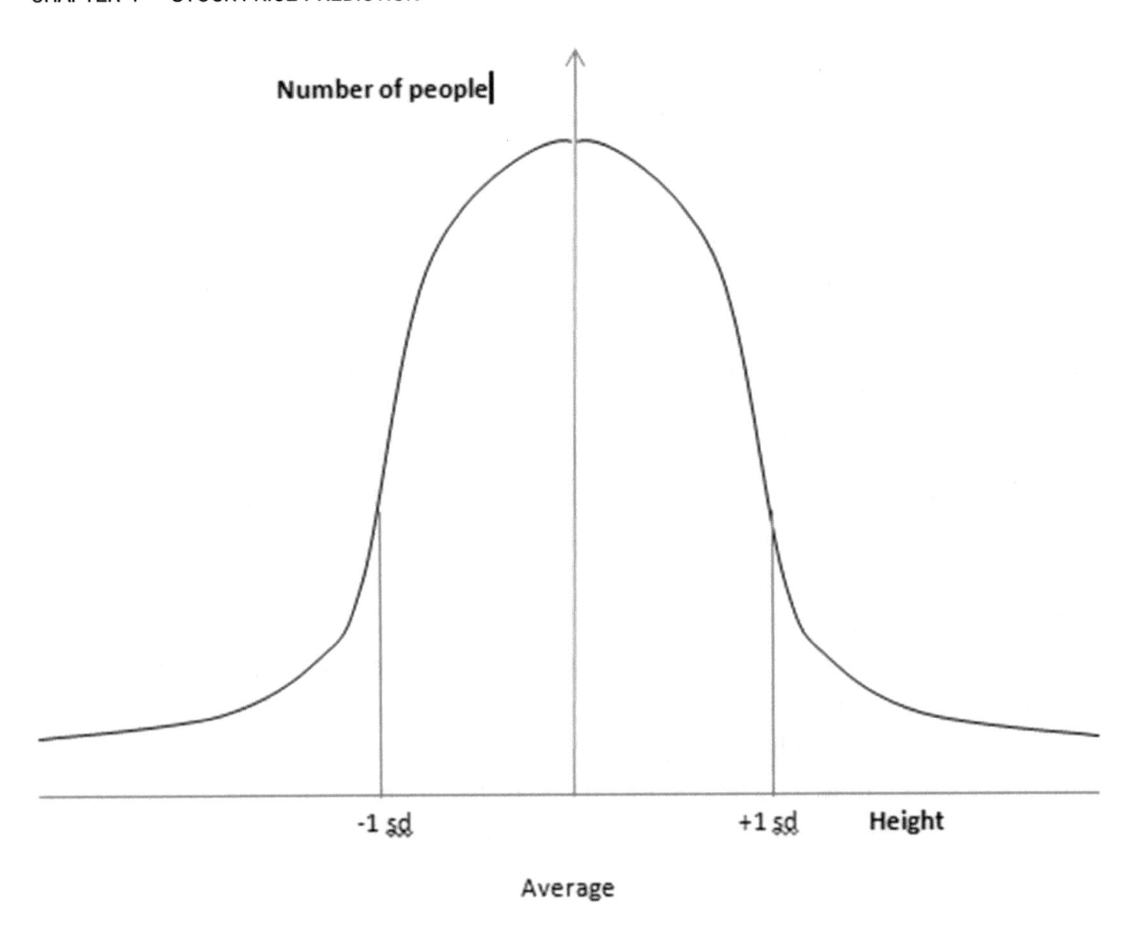

Figure 4-2. *Normal Distribution Function*

A more useful variation of the preceding Normal Distribution Function is the Cumulative Normal Distribution Function.

In the case of measuring the heights of people, we find the specific frequency of certain ranges and then add these up to find the total frequency within the two or more ranges. This mechanism is shown in the following diagram (Figure 4-3).

Height (feet – inches)	Frequency	Cumulative frequency
4 ft 6 in < h <= 4ft 9 in	1	1
4 ft 9 in < h <= 5ft 0 in	2	3 (2+1)
5 ft 0 in < h <= 5ft 3 in	4	7 (4+2+1)
5 ft 3 in < h <= 5ft 6 in	9	16 (9+4+2+1)
5 ft 6 in < h <= 5ft 9 in	25	41 (25+9+4+2+1)
5 ft 9 in < h <= 6ft 0 in	26	67 (26+5+9+4+2+1)
6 ft 0 in < h <= 6ft 3 in	17	84 (17+25+9+4+2+1)
6 ft 3 in < h <= 6ft 6 in	10	94 (10+17+25+9+4+2+1)
6 ft 6 in < h <= 6ft 9 in	6	100 (6+10+17+25+9+4+2+1)

Figure 4-3. *Cumulative distribution mechanism*

In the table we have the number of people in each of the ranges. If we want to know, for example, the number of people whose height is between 4 feet 6 inches and 5 feet, then this is just the sum of the first two ranges (1+2) which is 3. The total number in the range 4 feet 6 inches to 5 feet 3 inches is 1+2+4 which is 7.

When we plot the heights against the cumulative frequency, we get the following graph (Figure 4-4).

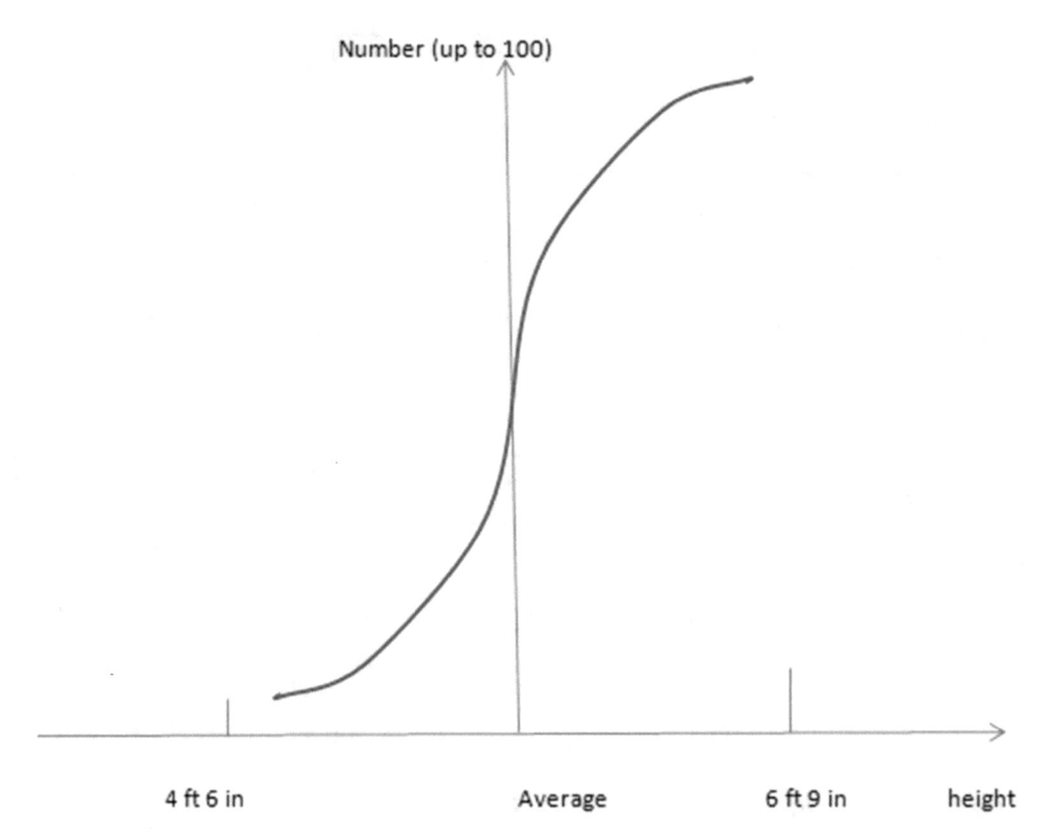

Figure 4-4. *Cumulative frequency vs. height graph*

For the purposes of finding an appropriate random value for our stock price formula, we use the Cumulative Normal Distribution Function diagram shown in Figure 4-5. In this case we see that the number from our diagram earlier can be converted to a probability where the probability of the height being below 6 ft. 10 inches (for our survey) is 1 and the probability of heights being below 4 ft. 6 inches is zero.

We will use this cumulative normal distribution curve along with our random number generator to find the x values.

For the random part of the formula, we need to use the Normal Distribution. If we made a note of the actual random daily fluctuations of the stock prices, we would find that the values, when plotted on a graph, would have a Normal Distribution. So that if we just used our random number generator, we would get an even chance that 0.0001 and 0.5000 would come up. This does not follow the Normal Distribution. So what we do is use our random number generator to give us our random numbers from 0 to 1 and use these as the Probability part of the Cumulative Normal Distribution graph (also 0 to 1) and then find, from the graph, the corresponding x value for this value of probability. We then know that our x values must follow a Normal Distribution and so it is consistent with the actual x values (in this case stock prices) would be.

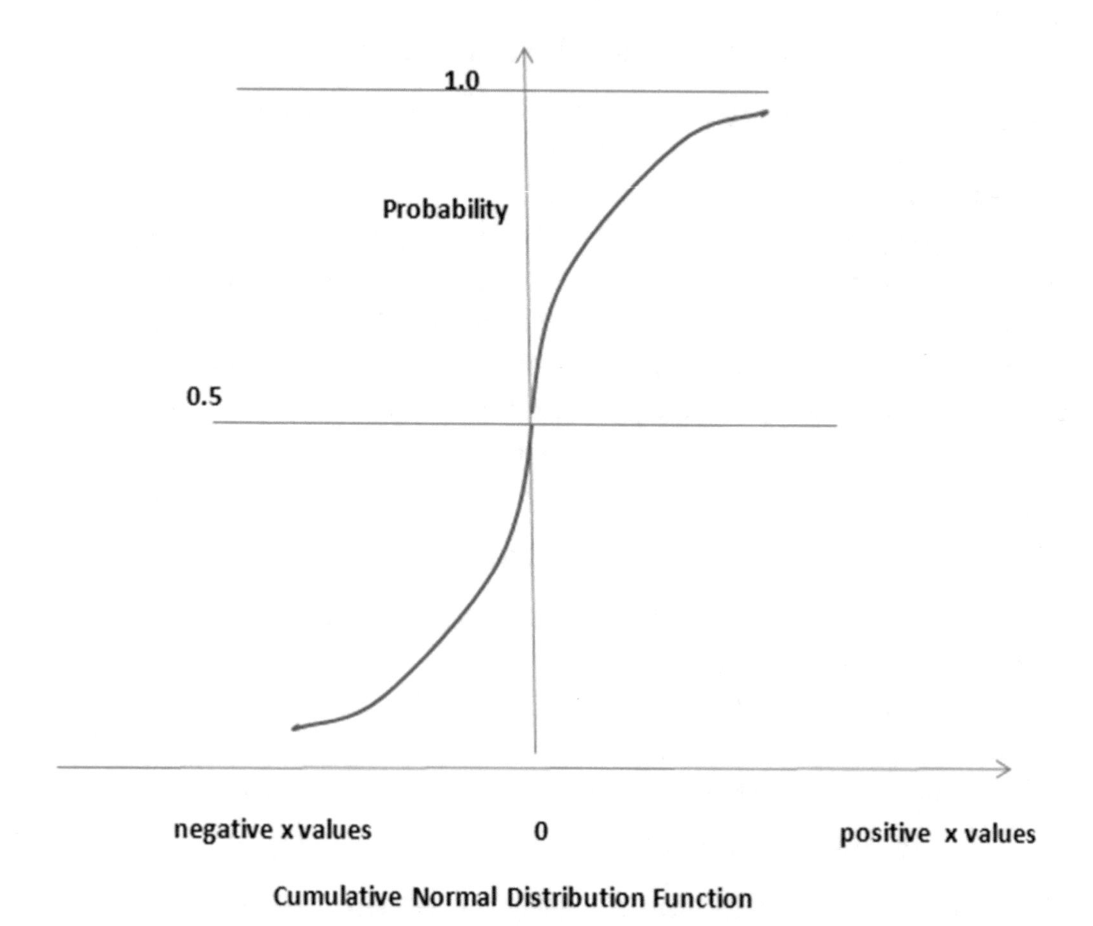

Figure 4-5. *Cumulative Normal Distribution Function*

Each point on the x axis is a point of cumulative frequency. Notice that the graph above 0.5 is symmetrical with that below 0.5. So when we calculate the x values above 0.5, we just take the equivalent value below 0.5 and multiply it by -1, that is, if we find a probability of 0.25 and, say, x = -1.7, then we take the probability as 1-0.25 (=0.75), so x is +1.7.

We want to use the graph to find these random x points as we know they form the normal distribution we want. If we can tell the 0-1 random number generated by our computer program as the probability value from the graph (also 0-1), then we can take the corresponding x value from the graph as our random value in our formula. Unfortunately, there is no mathematical formula to get the x value from a given probability value, but there is an algorithm to find it and we can code this algorithm into our program.

From the algorithm, the x value is given by

$$\mathbf{x = t - (c_0 + c_1 t + c_2 t^2) / (1 + b_1 t + b_2 t^2 + b_3 t^3)}$$

where $t = \sqrt{\ln(1 / q^2)}$ where q is our computer-generated random number

$c_0 = 2.515517$ $c_{1-} = 0.802853$ $c_2 = 0.010328$

$b_1 = 1.432788$ $b_2 = 0.189269$ $b_3 = 0.001308$

We can generate a random number between 0 and 1, and this represents the probability axis of the Cumulative Normal Distribution curve. Having the probability value of the curve, we can use the preceding algorithm to find the x value on the x axis.

From our graph, we know that a probability of 0.5 should give an x value of 0, so we can calculate the value of x from the preceding formula manually.

We can split the formula to make it easier to follow.

If we let ($\mathbf{c_0 + c_1 t + c_2 t^2}$) = c and ($\mathbf{1 + b_1 t + b_2 t^2 + b_3 t^3}$) = b, then the formula becomes

$$\mathbf{x = t - c / b} \ (1)$$

As our probability value is 0.5, then we assign this to q in the formula.

We get $t = \sqrt{\ln(1 / q^2)} = \sqrt{\ln(1 / 0.5)} = 1.177410$.

So $t^2 = (1.177410)^2 = 1.386294$.

And $t^3 = (1.177410)^3 = 1.632237$.

Substituting these values into the preceding formula (1) for x, we get

$$c = (2.51557) + (0.802853)*(1.177410) + (0.010328)*(1.386294)$$

$$b = 1+(1.432788)*(1.177410)+(0.189269)*(1.386294)+(0.001308)*(1.632237)$$

So c/b = 1.177410.

So our formula (1) becomes x = 1.177410 – 1.177410, or x = 0, which is what we want. From our Cumulative Normal Distribution graph earlier, when the probability is 0.5, x = 0.

We can write a short program to do the preceding calculation, and we can allow the user to enter the probability factor themselves. So if they entered 0.5, they would get x=0 as in our calculation.

The program prints the corresponding x value for the probability entered.

The code for this is shown as follows:

```
/* assetalgorithm.c */
/* Stock price predictor simulation */
/* tests inverse cumulative */
/* normal distribution function */
/* User enters a probability value */

#define _CRT_SECURE_NO_WARNINGS
#include <stdio.h>
#include <math.h>
#include <stdlib.h>
#include <time.h>

      double c,c0,c1,c2,d1,d2,d3;
      double q,d,x,y,t;

      int i;
      time_t tim;

      int n;

      double probvalue;

int main()
{

      /* Set values for cumulative normal distribution formula */
      c0=2.515517;
      c1=0.802853;
      c2=0.010328;
      d1=1.432788;
      d2=0.189269;
      d3=0.001308;

      /* User enters the probability used to find the x value */
      printf("Please enter Probability value between 0.0 and 1.0 :\n ");
      scanf("%lf", &probvalue);
      printf("You entered %lf\n", probvalue);
```

```
    if(probvalue < 0.0 || probvalue > 1.0)
          return; /* Entered value is out of range */

    y=probvalue; /* Store the entered value */

    /* Use the symmetry of the Cumulative Normal Distribution graph */
    if(y>=0.5)
          q=1-y;
    else
          q=y;

    /* Calculate the values in the formula */
    t=sqrt(log(1/pow(q,2)));
    c=c0+c1*t+c2*pow(t,2);
    d=1+d1*t+d2*pow(t,2)+d3*pow(t,3);

    x=t-(c/d);

    /* Use the symmetry of the Cumulative Normal Distribution graph */
    if(y < 0.5)
    {

           y=-1.0*x;

    }
    else if(y == 0.5)
           y=0;
    else
           y=x;

    /* Print the x result for the entered Probability value */
    printf("probvaluevalue  = %lf\n",probvalue);
    printf("x = %lf\n",y);

    return;

}
```

By entering different values for the probability, you can build up a picture of what x values you get for the probability values. You should get the distribution shown in Figure 4-5. There is an exercise you can do at the end of this chapter to demonstrate this.

4.5 Combining the Two Elements

So if we have a list of data for, say the last 19 days stock prices, then we have all of the data required by our formula:

Today's Stock Price = Yesterday's Stock price * exp(Drift + Random Change)

The Random Change part of this is our random number, as generated earlier, multiplied by the standard deviation as described earlier:

Random Change = PDRstd_deviation*calcrand

where **PDRstd_deviation** is found from the **PDRvariance** which we saw earlier was

$$\textbf{PDRvariance} = (\sum(\textbf{pdrx} - \textbf{pdr}\bar{\textbf{x}})^2) / \textbf{n}$$

And from our definition earlier of standard deviation (denoted by σ)

$$\sigma = \sqrt{(\textbf{variance})}$$

And **calcrand** is our calculated random number using the preceding algorithm.

We can read the previous 19 days stock prices or we can preset them in the program. In our following program, we have preset the values where the first value in the array is yesterday's price, the second value is the day before yesterday's price, and so on.

We have a function, `avstdvar`, which calculates the average values, standard deviation and variance. The function `calcrand` contains the algorithm mentioned earlier for calculating an appropriate random number.

We preset the historical stock prices into the array `dayvals`. The second part of this array will contain out predicted future prices when we have calculated them.

We work out the Periodic Daily Return from our historical stock prices using the formula

Periodic Daily Return = ln (Day's Price / Previous Day's Price)

where ln is the natural logarithm.

Note that if we have 19 historical day's prices, then we can only find 18 Periodic Day Returns because when we get to the 19th one, we don't have the previous day's return. In our following program, we have 19 historical day's prices, so we will have 18 Periodic Day Returns.

```
/* asseta2.c */
/* Stock price predictor simulation */
/* from Day values */

#define _CRT_SECURE_NO_WARNINGS
#include <stdio.h>
#include <math.h>
#include <stdlib.h>
#include <time.h>

double calcrand(); /* Function to calculate our random value for x using
the formula */
void avstdvar(double dayvals[]); /* Function to calculate variance and
standard dev */

      double c,c0,c1,c2,d1,d2,d3;
      double q,d,F,y,t;

      int i,j;
      time_t tim;

      int n;
      double average, variance, std_deviation, sum = 0.0, sum1 = 0.0;

      /* The PDR prefix to variables denotes Periodic Day Return */
      double PDRaverage,PDRvariance,PDRstd_deviation,pdrsum,nextval,
      lastval,drift,epsilon,exptest,nitest;
      double pdr[50];

void main()
{
      FILE *fp;
```

```
/* Array containing day stock prices starting with yesterday and
moving backward through previous days*/
/* The part of the array following these preset values will
contain our /*
/* calculated stock price values. So the whole array can be
printed out */
/* on our graph */
double dayvals[50]={22.82,22.51,22.47,22.05,22.96,21.43,20.97,20.46,
20.25,20.46,20.45,20.7,20.31,20.94,20.85,20.59,20.65,21.12,20.78};

double value,testval;
int j;

fp=fopen("asseta2.dat","w");
srand((unsigned) time(&tim)); /* Set up random number function */

for(i=0;i<50;i++)
{
      /* Clear predicted rate array */
      pdr[i]=0.0;
}
for(i=19;i<50;i++)
{
      /* Clear the end part of our values array for our predicted vales */
      dayvals[i]=0.0;

}

n=19; /* Number of historical day's prices */


j=0;
/* Write historical stock prices to output file */
/* Work backward through dayvals as */
/* the array is preset with today's in the first */
/* position, yesterday's in second position, */
/* and so on */
```

```
    for(i=18; i >-1; i--)
    {
        fprintf(fp,"%d\t%lf\n",j,dayvals[i]);
        j++;
    }

    /* Calc PDRs - if you enter 19 days there will be 18 PDRs */


    for(j=0;j<n-1;j++)
    {
        pdr[j]=log(dayvals[j]/dayvals[j+1]);

    }
    /* Compute the sum of all PDR elements */
    /* Find PDR average */
    pdrsum=0.0;
    for (i = 0; i < n-1; i++)
    {
        pdrsum = pdrsum + pdr[i];
    }
    PDRaverage = pdrsum / (double)(n-1);

    /* Call function to calculate statistical values */
    avstdvar(dayvals);

    /* Calculate drift */
    drift=PDRaverage-(PDRvariance/2);

    lastval=dayvals[0];


    /* Calculate values using formula */
/*Today's Stock Price = Yesterday's Stock price * exp( Drift + Random Change)*/
/* we use the variable nextval for Today's Stock Price */
/* and the variable lastval for = Yesterday's Stock price */
/* and PDRstd_deviation*calcrand() for Random Change */
```

```
    /* nextval=lastval*exp(drift+PDRstd_deviation*calcrand()) */
    for (i = 19; i < 38; i++)
    {

          nitest=calcrand();
          exptest=exp(drift+PDRstd_deviation*nitest);
          nextval=lastval*exptest;

          fprintf(fp,"%d\t%lf\n",i,nextval);

          lastval=nextval; /* Set last value for the next iteration */

    }

    fclose(fp);

}
double calcrand()
{
/* Function to calculate our random value for x using the formula */
/* x = ( t - c0 + c1t + c2t²) / (1 + d1 + d2t² + d3t³) */

/* Set values for cumulative normal distribution formula */


    c0=2.515517;
    c1=0.802853;
    c2=0.010328;
    d1=1.432788;
    d2=0.189269;
    d3=0.001308;

    y=rand()%1000; /* Generate random number between 0 and 1 */
    y=y/1000;

    /* Use the symmetry of the Cumulative Normal Distribution graph */
    if(y>=0.5)
          q=1-y;
    else
          q=y;
```

```
    /* Apply the Cumulative Normal Distribution Algorithm */

    t=sqrt(log(1/pow(q,2)));
    c=c0+c1*t+c2*pow(t,2);
    d=1+d1*t+d2*pow(t,2)+d3*pow(t,3);

    F=t-(c/d);

    /* Use the symmetry of the Cumulative Normal Distribution graph */
    if(y < 0.5)
    {
          y=-1.0*F;
    }
    else if(y == 0.5)
          y=0;
    else
          y=F;

    return y;

}

void avstdvar(double dayvals[])
{
/* Function to calculate variance and standard deviation and average */

    /* Average, standard deviation, variance processing */
    sum = 0.0;
    sum1 = 0.0;

    /* Compute the sum of all dayvals elements */
    for (i = 0; i < n; i++)
    {
          sum = sum + dayvals[i];
    }
    average = sum / (double)n;

    /* Compute variance and standard deviation */
    for (i = 0; i < n; i++)
```

```
{
        sum1 = sum1 + pow((dayvals[i] - average), 2);
}
variance = sum1 / (double)n;

/* Compute PDRvariance and PDRstandard deviation */
sum1=0.0;
for (i = 0; i < n-1; i++)
{
        sum1 = sum1 + pow((pdr[i] - PDRaverage), 2);
}
PDRvariance = sum1 / (double)(n-1);

std_deviation = sqrt(variance);
PDRstd_deviation = sqrt(PDRvariance);

}
```

The data produced by the program is shown in Figure 4-6. The points to the left of the black vertical line (x=19) are those from the historical data we preset in the program. The points to the right of the line are those generated by the program as predicted daily stock prices.

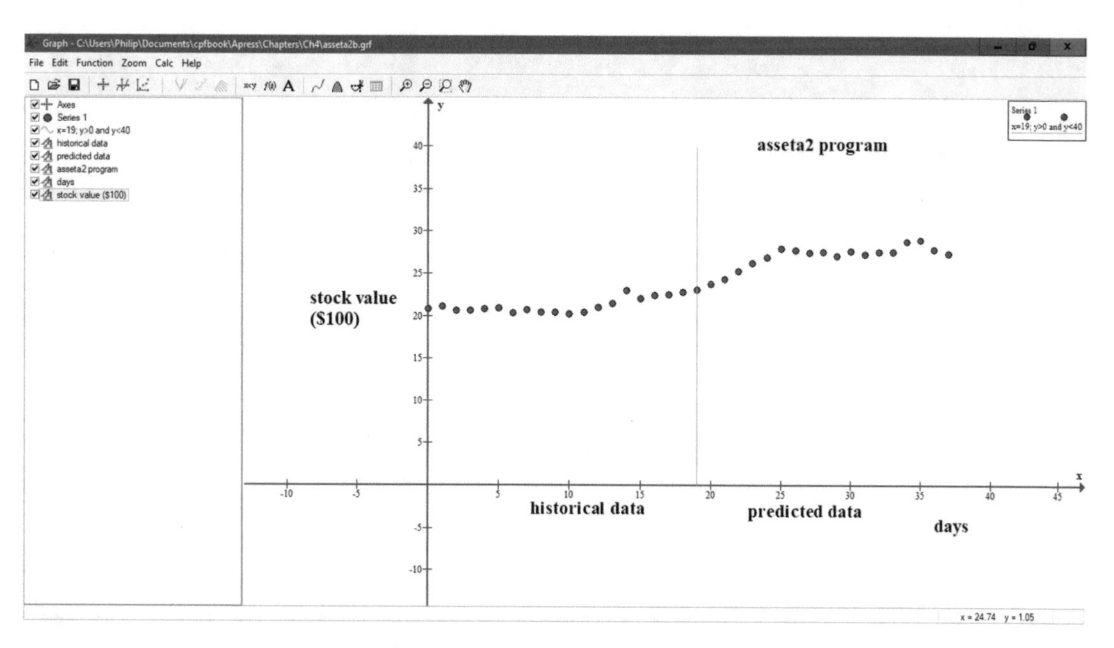

Figure 4-6. *Stock price prediction from historical data*

EXERCISES

1. Amend the asseta2.c program to read in user-entered historical values rather than using preset data.

   Read in the following data: 15.82,15.51,15.47,16.05,15.96,16.43,15.97,16.46, 17.25,17.46,17.45,17.7,17.31,17.94,17.85,17.59,18.65,19.12,18.78

   Print the graph. What do you notice about this graph compared to the one in the chapter?

2. Extend your **assetalgorithm.c** program from Section 4.3 of this chapter so that instead of entering probabilities yourself between 0 and 1, you generate random numbers between 0 and 1. If you have all of the calculations for one value of the probability to produce one x value, you can form this inside a forloop and write each value of x and the probability to a file. You can then display the file using graph and check the distribution.

PART II

Commercial Applications

CHAPTER 5

Supermarket Stock

5.1 What We Are Simulating

This chapter simulates how a supermarket keeps a check on its stock and when to order replacement stock from suppliers.

We create a file for the stock of different types of cheese that the supermarket sells. Each time a block of cheese goes through the checkout, a message is sent to the monitoring program which checks if a reorder is needed. If so, it outputs a message to say the type of cheese needed and the address where you need to send the order to. If a new supply is received by the supermarket, then the monitoring program is told this and the file is updated appropriately. This is a very simplified mechanism to illustrate the sequence of operations that are needed.

The file is made up by the items shown in the following.

The first number is the ID of the item. The second is its description (the type of cheese, which is Brie for ID 4). The third item is the limit point where, if the stock level gets down to this number, we will make a reorder. The fourth item is the current level of stock. The final item is the address where we need to send the order.

```
{4,"Brie      ",23,50,"95,West Park St"};
{7,"Gouda     ",34,51,"2,North Park St"};
{9,"Edam      ",44,52,"17,New Gate St"};
{11,"Camembert ",25,53,"12,Toll Av"};
{14,"Cheshire  ",34,54,"5,State Rd"};
{16,"Cheddar   ",51,55,"63,Madison St"};
{17,"Pecorino  ",23,56,"12,East Park St"};
{19,"Manchego  ",44,57,"14,May St"};
{23,"Provolone ",35,58,"20,Oregon Way"};
{24,"Parmigiano",40,59,"10,Park St"};
```


P. Joyce, *Practical Numerical C Programming*, https://doi.org/10.1007/978-1-4842-6128-6_5

```
{27,"Mascarpone",40,60,"31,Queen St"};
{31,"Mozzarella",42,61,"19,Hope Av"};
{32,"Feta      ",45,62,"13,Charles Av"};
{35,"Gruyere   ",47,63,"54,Tower St"};
{38,"Monterey  ",41,63,"11,Cardew Av"};
{44,"Gorgonzola",54,68,"26,Jones St"};
{47,"Stilton   ",58,69,"57,Lower St"};
```

The structure in the program for each item as described earlier is as follows:

```
struct super {
   int ID;
   char desc[11];
   int limit;
   int numstock;
   char address[30];
};
```

The following program creates the file as described earlier.

The file is created, then closed, and reread so that it can be printed out.

The code to set up this file is shown in the following:

```
/* createmarket.c  */
/* Creates supermarket stock file */
/* Prints out the records sequentially */
#define _CRT_SECURE_NO_WARNINGS
#include<stdio.h>
 /* Structure for each item in stock.*/
/* The "desc" part is the name of a type of cheese */
/* "limit" is the minimum level of stock */
/* after which a reorder is required */
/* "numstock is the current level of the item */
/* "address" is where to get the order */
struct super {
   int ID;
   char desc[11];
   int limit;
```

```
   int numstock;
   char address[30];
};

void main()
{
      int i,numread;
      FILE *fp;
      struct super s1;
      struct super s2;

      /* Preset structures for 17 types of cheese in our file */
      struct super s10 = {4,"Brie       ",23,50,"95,West Park St"};
      struct super s11 = {7,"Gouda      ",34,51,"2,North Park St"};
      struct super s12 = {9,"Edam       ",44,52,"17,New Gate St"};
      struct super s13 = {11,"Camembert ",25,53,"12,Toll Av"};
      struct super s14 = {14,"Cheshire  ",34,54,"5,State Rd"};
      struct super s15 = {16,"Cheddar   ",51,55,"63,Madison St"};
      struct super s16 = {17,"Pecorino  ",23,56,"12,East Park St"};
      struct super s17 = {19,"Manchego  ",44,57,"14,May St"};
      struct super s18 = {23,"Provolone ",35,58,"20,Oregon Way"};
      struct super s19 = {24,"Parmigiano",40,59,"10,Park St"};
      struct super s20 = {27,"Mascarpone",40,60,"31,Queen St"};
      struct super s21 = {31,"Mozzarella",42,61,"19,Hope Av"};
      struct super s22 = {32,"Feta      ",45,62,"13,Charles Av"};
      struct super s23 = {35,"Gruyere   ",47,63,"54,Tower St"};
      struct super s24 = {38,"Monterey  ",41,63,"11,Cardew Av"};

      struct super s28 = {44,"Gorgonzola",54,68,"26,Jones St"};
      struct super s29 = {47,"Stilton   ",58,69,"57,Lower St"};

      /* Open the supermarket file */
      fp = fopen("superm.dat", "w");
      /* Write details of each structure to file*/
      /* From the structures defined earlier */
```

```
fwrite(&s10, sizeof(s1), 1, fp);
fwrite(&s11, sizeof(s1), 1, fp);
fwrite(&s12, sizeof(s1), 1, fp);
fwrite(&s13, sizeof(s1), 1, fp);
fwrite(&s14, sizeof(s1), 1, fp);
fwrite(&s15, sizeof(s1), 1, fp);
fwrite(&s16, sizeof(s1), 1, fp);
fwrite(&s17, sizeof(s1), 1, fp);
fwrite(&s18, sizeof(s1), 1, fp);
fwrite(&s19, sizeof(s1), 1, fp);
fwrite(&s20, sizeof(s1), 1, fp);
fwrite(&s21, sizeof(s1), 1, fp);
fwrite(&s22, sizeof(s1), 1, fp);
fwrite(&s23, sizeof(s1), 1, fp);
fwrite(&s24, sizeof(s1), 1, fp);
fwrite(&s28, sizeof(s1), 1, fp);
fwrite(&s29, sizeof(s1), 1, fp);
/* Close the file */
fclose(fp);
/* Reopen the file */
fp=fopen("superm.dat", "r");
/* Read and print out all of the records on the file */
printf("\nID DESCRIPTION LIMIT NUMBER IN STOCK ADDRESS");
for(i=0;i<17;i++)
{
      numread=fread(&s2, sizeof(s2), 1, fp);
      if(numread == 1)
      {
            printf("\n%2d : %s : %d : %d : %s", s2.ID,s2.desc,s2.
            limit,s2.numstock,s2.address); /* Note the 2d as we want
            2 digits */
      }
```

```
        else {
                /* If an error occurred on read, then print out message */
                        if (feof(fp))
                        printf("Error reading superm.dat : unexpected end
                        of file fp is %p\n",fp);
                        else if (ferror(fp))
                {
                        perror("Error reading superm.dat");
                }
        }
    }
    /* Close the file */
    fclose(fp);
}
```

5.2 Updating the File

Once we have the file created, we can write our updating program. In order to simulate what happens in a real supermarket (which will use barcodes when the item goes through the checkout), here we will type in the ID and how many items are being put through the checkout.

We first read the file and write it to the screen. This is not required at the actual checkout but, again, it makes our simulation easier to follow.

The two main operations of the program are a stock update, where the supermarket has received new stock from its suppliers, and an item being sold through the checkout. So the program asks the user which of these two processes is being done.

If it is a new supply of stock, then the user is asked to enter the stock's ID and the amount of stock to be added. A function is then called to do the update. This is done using the instruction `updatefunc(stockitemID,updateamount);`. This passes to the function the ID and the amount to be updated. The function itself is defined at the end of the program. It reads through the file until it finds the correct ID. At this stage the file pointer is pointing to the next record in the file, so we have to back up to the previous record using the instruction `fseek(fp,minusone*sizeof(s2),SEEK_CUR);`.

To explain this mechanism of backup, look at the following situation.

We are reading through the file and we have just read the second entry in the file, so the file pointer is now pointing to the third entry as shown in the following:

```
4       Brie        23    50
7       Gouda       34    51
9       Edam        44    52     ← file pointer
11      Camembert   25    53
14      Cheshire    34    54
```

The fread function `fread(&s2, sizeof(s2), 1, fp);` will have placed the details for the second structure (Gouda) into the structure variable s2. If we now want to update the number of items in stock of Gouda from 51 to 49 in s2 and we want to write this back to the file, we need to move the pointer back to pointing at the Gouda structure of the file. We do this using `fseek(fp,minusone*sizeof(s2),SEEK_CUR);`.

The `minusone*sizeof(s2)` part of the instruction tells the file pointer to move back by the length of the s2 structure. After the instruction, the pointer will be pointing at the Gouda file entry as shown in the following:

```
4       Brie        23    50
7       Gouda       34    51     ← file pointer
9       Edam        44    52
11      Camembert   25    53
14      Cheshire    34    54
```

Now we can write the new data to the file using `fwrite(&s2, sizeof(s2), 1, fp);`. The fwrite instruction writes the new contents of s2 to the position in the file pointed to. The new data is then printed to the user. This completes the updating mechanism of the program, and the number for the stock of Gouda is updated to 49 as shown. After the `fwrite`, the pointer will be pointing to the next item.

```
4       Brie        23    50
7       Gouda       34    49
9       Edam        44    52     ← file pointer
10      Camembert   25    53
14      Cheshire    34    54
```

The other mechanism that the program performs is when an item is sold. The operation for this is similar to the new stock mechanism. In this case the `decreasefunc(stockitemID,updateamount); function` is called. In this function the ID is entered and the number of items for this ID that have been sold. The ID is found in the file and the number of items entered is subtracted from the current level on the file. A test is then made to see if this new level is equal to or below the minimum level when a reorder is needed. If this level has been reached, the reorder() function is called which outputs a message to the user. The new level is then written to the file using the `fseek(fp,minusone*sizeof(s2),SEEK_CUR);` mechanism as described earlier.

The file is then closed.

The code for updating the file is shown in the following. The user is asked to enter the ID for the stock item being updated. Then they are asked if they want to increase or decrease the stock. If they say they want to increase the stock, then a function called `updatefunc();` is called. If they say they want to decrease the stock, then a function called `decreasefunc();` is called.

The code is shown as follows:

```
/* markettest4.c */
/* Supermarket reordering simulation */
#define _CRT_SECURE_NO_WARNINGS
#include <stdio.h>
#include <math.h>
#include <stdlib.h>
#include <time.h>

/* Functions defined at the end of the program */
void reorder(); /* Call a reorder function */
void updatefunc(); /* Increase stock function */
void decreasefunc(); /* Decrease stock function */

/* structure for each item in stock.*/
/* "ID" is the ID for the item */
/* The "desc" part is the name of a type of cheese */
/* "limit" is the minimum level of stock */
/* after which a reorder is required */
/* "numstock" is the current level of the item */
/* "address" is where to get the order */
```

```
struct super {
   int ID;
   char desc[11];
   int limit;
   int numstock;
   char address[30];
   };

struct super s1;
struct super s2;
struct super st[17];
FILE *fp;
long int minusone = -1;
int i;

int main()
{

      int numread;
      int stockitemID,updateamount;
      char update;

      /* Open the supermarket file */
      fp=fopen("superm.dat", "r");

      /* Read and print out all of the records on the file */
      printf("\nID DESCRIPTION LIMIT NUMBER IN STOCK ADDRESS");
      for(i=0;i<17;i++)
      {
            numread=fread(&s2, sizeof(s2), 1, fp);

            if(numread == 1)
            {
                  printf("\n%2d : %s : %d : %d : %s", s2.ID,s2.desc,s2.
                  limit,s2.numstock,s2.address); /* note the 2d as we want
                  2 digits */
            }
```

```
        else {
                /* If an error occurred on read, then print out message */

                        if (feof(fp))

                        printf("Error reading superm.dat : unexpected end
                        of file fp is %p\n",fp);

                        else if (ferror(fp))
                {
                        perror("Error reading superm.dat");
                }
        }

}
/* Close the file */

fclose(fp);

/* Ask the user what they want to do with the file */
/* Increase or decrease the stock? */

printf("\nIs this a Stock update(increase) ? (y or n) \n");
scanf("%c", &update);
printf("\n answer is %c\n",update);
if(update == 'y')
{
        /* User wants to update(increase) stock level */
        printf("\nenter ID \n");
        scanf("%d", &stockitemID);

        printf("\n ID is %d",stockitemID);

        printf("\nenter update amount \n");
        scanf("%d", &updateamount);

        /* Call function to update the stock */
        updatefunc(stockitemID,updateamount);

        return;
}
```

```
        else if(update == 'n')
        {
                /* User wants to decrease stock level */
                printf("\nenter ID \n");
                scanf("%d", &stockitemID);

                printf("\n ID is %d",stockitemID);

                printf("\nenter number sold \n");
                scanf("%d", &updateamount);

                printf("\n number sold entered is %d",updateamount);

                /* Call function to decrease the stock */
                decreasefunc(stockitemID,updateamount);
                fclose(fp);

                return;
        }

}
void reorder()
{
        /* Function to say that you have reordered and the address */
        printf("\nreorder called");
        printf("\n address is %s",s2.address);
}
void updatefunc(int stockitemID,int updateamount)
{

        /* Increase stock function */

        /* Function to update current level of stock */
        /* After you to a read, the file pointer will be */
        /* pointing to the next record in the file */
        /* So we move the file pointer backward to */
        /* point to the record we have just read */
        /* using the fseek command */
```

```
        printf("\nupdate called");
        printf("\nstockitemID is %d updateamount is %d\n",stockitemID,
        updateamount);
        fp = fopen("superm.dat", "r+");
        for (i = 0;i < 17;i++)
        {
                /* Read each pressure data from file sequentially */

                fread(&s2, sizeof(s2), 1, fp);

                if(s2.ID == stockitemID)
                {
                        /* We have found the one we want to update */
                        s2.numstock = s2.numstock + updateamount;

                        fseek(fp,minusone*sizeof(s2),SEEK_CUR);

                        fwrite(&s2, sizeof(s2), 1, fp);

                        printf("\n ID is %d",s2.ID);
                        printf("\n limit is %d",s2.limit);
                        printf("\n numstock is %d",s2.numstock);
                        printf("\n address is %s",s2.address);

                        fclose(fp);
                        break;
                }
        }
}
void decreasefunc(int stockitemID,int updateamount)
{

                /* Decrease stock function */

                /* After you to a read, the file pointer will be */
                /* pointing to the next record in the file */
                /* So we move the file pointer backward to */
                /* point to the record we have just read */
                /* using the fseek command */

                /* Open supermarket file */
```

```
fp = fopen("superm.dat", "r+");
for (i = 0;i < 17;i++)
{
      fread(&s2, sizeof(s2), 1, fp);
      if(s2.ID == stockitemID)
      {
            st[i].ID = s2.ID;
            st[i].limit = s2.limit;
            st[i].numstock = s2.numstock;
            if(st[i].numstock == 0)
            {
                  printf("\n Out of stock");
                  printf("\n numstock is %d",st[i].numstock);
                  printf("\n limit is %d",s2.limit);
                  printf("\n number sold is %d",updateamount);
                  break;
            }
            if(st[i].numstock - updateamount <= 0)
            {
                  /*After decrease, stock level is zero or
                  below*/
                  printf("\nStock update");
                  st[i].numstock = 0; /* set to zero (negative
                  is impossible) */
                  s2.numstock = st[i].numstock;
                  reorder();
            }
```

```
                    else if(st[i].numstock  - updateamount <= s2.limit)
                    {
                            /*After decrease, stock level is below
                            limit*/
                            printf("\nStock update");
                            s2.numstock = st[i].numstock-updateamount;

                            reorder();
                    }
                    else
                    {
                            /*After decrease, stock level is above
                            limit*/
                            printf("\nStock update");
                            st[i].numstock = st[i].numstock -
                            updateamount;
                            s2.numstock = st[i].numstock;

                    }
                    printf("\n limit is %d",s2.limit);
                    printf("\n number sold is %d",updateamount);
                    printf("\n numstock is %d",s2.numstock);

                    /* Move the file pointer back by one record */
                    fseek(fp,minusone*sizeof(s2),SEEK_CUR);

                    fwrite(&s2, sizeof(s2), 1, fp);

                    break;
            }
        }
}
```

The updated values are printed at the end of the program. Another short program could be written to print out the current values of the file just in case the user has lost track of what has happened previously. This program is given as an exercise.

EXERCISES

1. Write a program to read the file and output to the user. This is useful if the user needs to check the current stock.

CHAPTER 6

Flight Information

6.1 Airport Display Boards

This chapter shows the mechanism for displaying flight information on airport display boards. Here, we have only looked at flight arrivals but a similar mechanism can be used for flight departures. As with the previous chapter, the mechanisms used here have been modified so that the user can control what is happening.

6.2 Create Flights File

What we will do is have a file containing 17 flights. We will assume that each display board will only display 12 flights. These will be the "current" flights, those which are relevant to the current time of day. So at the start of the day, these 12 flights will be the first 12 flights in the file (as the flights are in chronological order). In reality, depending on the airport, there could be maybe 200 flights in the file, but the boards would still just show the current 12.

The data for the 17 flights we will be using is shown in the following.

The data for each flight is kept as a separate record on the arrivals.dat file.

The first item for each record in the file is the position on the file of the flight. So here, they are numbered 1 to 17. The second item is the flight code. Here, for the first flight, it is AA1232. The next two fields are the scheduled arrival time of the flight and the expected arrival time. For the first two flights, these are the same. The next field is the city of origin of the flight.

```
{"1","AA1232","07:00","07:00","CHICAGO",""};
{"2","BA123","07:05","07:05","LONDON",""};
{"3","AA4517","07:08","07:15","BOSTON",""};
```


P. Joyce, *Practical Numerical C Programming*, https://doi.org/10.1007/978-1-4842-6128-6_6

```
{"4","AF123","07:10","07:10","PARIS",""};
{"5","NH444","07:20","07:20","TOKYO",""};
{"6","DJ144","07:22","07:22","MUMBAI",""};
{"7","AZ2348","07:23","07:25","WASHINGTON",""};
{"8","VS9745","07:25","07:26","TORONTO",""};
{"9","DL5816","07:30","07:30","CHICAGO",""};
{"10","KL5393","07:33","07:33","MANCHESTER",""};
{"11","AZ4627","07:35","07:40","ROME",""};
{"12","VS4677","07:40","07:40","NEW ORLEANS",""};
{"13","SQ125","07:45","07:45","FRANKFURT",""};
{"14","EI5666","07:48","07:48","LONDON",""};
{"15","WS2321","07:50","07:50","DULLES",""};
{"16","AA197","07:55","08:00","SAN FRANCISCO",""};
{"17","B57321","07:58","07:48","SARASOTA",""};
```

This flight data is preset in structures in the program and these are written to the file. Each structure described earlier is held in the following structure:

```
struct arrivals {
        char posn[3];
        char flight_no[8];
        char sch_arrival_time[6];
        char exp_arrival_time[6];
        char origin[15];
        char remarks[14];
};
```

In addition to the fields described earlier, there is a remarks field. This is used, for example, when the aircraft is approaching the airport so the user can enter “Approaching”. When the aircraft has landed, you would see “Landed” on the boards.

When the aircraft has landed and all of the passengers have collected their luggage, the flight can be rolled off the board. It can be deleted from the file. In order to keep a running total of how many flights are on the file, we have a file called flightcnt.dat.

This file is created at the start of the program. The structure associated with the file is shown as follows:

```
struct flightcount {
        int count;
};
```

The structure is preset to its initial value of 17 (for our start of day 17 flights) as shown as follows:

```
struct flightcount fc={17};
```

So after creation of the file, we write this structure to the file and close it.

This file is used by the monitoring program and can also be inspected by the user by calling the program `fltcnt` (given as an exercise at the end of the chapter).

The code to create the arrivals.dat file is shown as follows:

```
/* createflightsb.c */
/* Creates file */
/* Prints out the records sequentially */

#define _CRT_SECURE_NO_WARNINGS
#include<stdio.h>
#include <string.h>
/* Arrivals flights structure */
 struct arrivals {
      char posn[3];
      char flight_no[8];
      char sch_arrival_time[6];
      char exp_arrival_time[6];
      char origin[15];
      char remarks[14];
};

struct flightcount {
      int count;
};
void main()
{
      int i,numread;
      FILE *fparr;
      FILE *fltcnt;
```

```
struct arrivals s1;
struct flightcount fc={17};
struct flightcount fcr;

/* Preset individual structures with data for each flight */
struct arrivals s10 = {"1","AA1232","07:00","07:00","CHICAGO",""};
struct arrivals s11 = {"2","BA123","07:05","07:05","LONDON",""};
struct arrivals s12 = {"3","AA4517","07:08","07:15","BOSTON",""};
struct arrivals s13 = {"4","AF123","07:10","07:10","PARIS",""};
struct arrivals s14 = {"5","NH444","07:20","07:20","TOKYO",""};
struct arrivals s15 = {"6","DJ144","07:22","07:22","MUMBAI",""};
struct arrivals s16 = {"7","AZ2348","07:23","07:25","WASHINGTON",""};
struct arrivals s17 = {"8","VS9745","07:25","07:26","TORONTO",""};
struct arrivals s18 = {"9","DL5816","07:30","07:30","CHICAGO",""};
struct arrivals s19 = {"10","KL5393","07:33","07:33","MANCHESTER",""};
struct arrivals s20 = {"11","AZ4627","07:35","07:40","ROME",""};
struct arrivals s21 = {"12","VS4677","07:40","07:40","NEW ORLEANS",""};
struct arrivals s22 = {"13","SQ125","07:45","07:45","FRANKFURT",""};
struct arrivals s23 = {"14","EI5666","07:48","07:48","LONDON",""};
struct arrivals s24 = {"15","WS2321","07:50","07:50","DULLES",""};
struct arrivals s25 = {"16","AA197","07:55","08:00","SAN FRANCISCO",""};
struct arrivals s26 = {"17","B57321","07:58","07:48","SARASOTA",""};

/* Create the file flightcnt.dat which will contain */
/* the current number of flights in arrivals.dat. */
/* This file can then be updated when flights are */
/* removed from arrivals.dat to keep a running total */
fltcnt = fopen("flightcnt.dat","w");
fwrite(&fc, sizeof(fc), 1, fltcnt);
fclose(fltcnt);

fltcnt = fopen("flightcnt.dat","r");
fread(&fcr, sizeof(fcr), 1, fltcnt);
printf(" Number of flights : %d", fcr.count);
fclose(fltcnt);
/* Open the arrivals file */

fparr = fopen("arrivals.dat", "w");
```

```
/* Write details of each flight to file*/
/* from the structures defined earlier */

fwrite(&s10, sizeof(s1), 1, fparr);
fwrite(&s11, sizeof(s1), 1, fparr);
fwrite(&s12, sizeof(s1), 1, fparr);
fwrite(&s13, sizeof(s1), 1, fparr);
fwrite(&s14, sizeof(s1), 1, fparr);
fwrite(&s15, sizeof(s1), 1, fparr);
fwrite(&s16, sizeof(s1), 1, fparr);
fwrite(&s17, sizeof(s1), 1, fparr);
fwrite(&s18, sizeof(s1), 1, fparr);
fwrite(&s19, sizeof(s1), 1, fparr);
fwrite(&s20, sizeof(s1), 1, fparr);
fwrite(&s21, sizeof(s1), 1, fparr);
fwrite(&s22, sizeof(s1), 1, fparr);
fwrite(&s23, sizeof(s1), 1, fparr);
fwrite(&s24, sizeof(s1), 1, fparr);

fwrite(&s25, sizeof(s1), 1, fparr);
fwrite(&s26, sizeof(s1), 1, fparr);

/* Close the file */

fclose(fparr);
/* Reopen the file */

fopen("arrivals.dat", "r");

/* Read and print out all of the records on the file */
printf("\n Flight :Sched: Exp: Origin Remarks");
for(i=0;i<17;i++)
{
      numread=fread(&s1, sizeof(s1), 1, fparr);
      if(numread == 1)
      {
```

```
                printf("\n :%s\t%s\t%s\t%s\t%s\t%s", s1.posn,s1.flight_
                no,s1.sch_arrival_time,s1.exp_arrival_time,s1.origin,s1.
                remarks);
            }
            else {
                /* If an error occurred on read, then print out message */

                    if (feof(fparr))

                    printf("Error reading arrivals.dat : unexpected
                    end of file fparr is %p\n",fparr);

                    else if (ferror(fparr))
                {
                    perror("Error reading arrivals.dat");
                }
            }
        }
        /* Close the file */
        fclose(fparr);
}
```

Each flight is set up in its own structure and then each structure is written to the file using `fwrite`.

The file is then closed and reopened and then each record is read and displayed to the user.

6.3 Update Display Boards

The program to update the boards is split into two sections. One updates data on the boards, for example, if the flight is delayed, the user can change the expected arrival time and possibly put a remark "Delayed" on the board. The other section of the program deals with "rollup" when a flight has landed and been on the ground for a certain length of time so that it can be removed (or rolled up) from the board.

To make this clear in the program, the two sections are called:

```
/* "ROLL UP" SECTION OF THE PROGRAM */
```

and

```
/* "NOT ROLL UP" SECTION OF THE PROGRAM */
```

The program has the same structure for each flight as the `createflights` program.

It starts by opening and reading the flightcnt.dat file and printing the number of flights to the user. The program then reads the file, arrivals.dat, created by the `createflights` program and prints out to the user the 12 flights currently on display on the boards.

The number 12 is chosen as the number of flights that the display board can hold. In the program there is code as follows:

```
if(opt2 < 12)
                    lim = opt2;
            else
                    lim = 12;
```

where the field opt2 is updated with the number of flights in the file from the flightcnt.dat file. Initially, there are 17 flights on the file, but when the number being rolled up means that there are fewer than 12 flights in the file, then we use this number rather than 12. So toward the end of the day, the number on the board will drop to 11, 10, 9, and so on.

The program then asks the user if they want to do a rollup. If they say "no," then they are asked which flight they want to amend and the item of data for the flight they want to amend.

6.3.1 Not-Rollup Mechanism

The user enters the flight number as displayed on the boards and is given the option of a number to type in. Here "1" means that they want to amend the scheduled arrival time, "2" means they want to change the expected arrival time, "3" means they want to change the airport of origin, and "4" means they want to add a remark. The program uses a switch based on the number entered to go to the appropriate code to perform the required update.

There is a structure called `st[17]` in the program. This holds the data as specified in the arrivals structure for each flight.

When the user enters which field they wish to update, it is this `st` structure which is updated. When this process is completed, the arrivals.dat file is closed and then opened again using `fparr = fopen("arrivals.dat", "w");`. This creates a new arrivals.dat file as we are using the "w" parameter in the `fopen` function call.

After this, the `st` structures for each flight are written back to the arrivals.dat file.

6.3.2 Rollup Mechanism

For the rollup mechanism, the program reads in from the user which flight is to be removed from arrivals.dat. It then reads through the file and writes all of the flight details to the file arrivals2.dat except for the flight which is to be removed. When it gets to this flight, the program updates the flightcnt.dat file with the changed count and does not write the flight details to arrivals2.dat.

At the end of this process, arrivals2.dat contains the updated flight details, so the arrivals.dat file is deleted and arrivals2.dat is renamed arrivals.dat and becomes the new arrivals file. The code for this mechanism is shown in the following. The arrivals2 file is opened and its address is stored in source. The arrivals file is opened and its address is stored in target. The while loop goes through the arrivals2 file and reads each character at a time using `fgetc(source)` and then writes the character to the arrivals file using `fputc(ch, target);`. The two files are then closed.

```
/* Copy arrivals2.dat to new arrivals.dat */
            remove("arrivals.dat");

            source = fopen("arrivals2.dat", "r");
            target = fopen("arrivals.dat", "w");

            while ((ch = fgetc(source)) != EOF)
                  fputc(ch, target);

            fclose(source);
            fclose(target);
```

This is then printed out to the user.

The code for this is shown as follows:

```
/* flightsg.c */
/* Airport display boards updates*/
```

```
/* arrivals only */
#define _CRT_SECURE_NO_WARNINGS
#include <stdio.h>
#include <math.h>
#include <stdlib.h>
#include <time.h>
#include <string.h>

/* Functions defined at the end of the program */

void notrollupfunc(); /* Not rollup function */
void rollupfunc(); /* Rollup function */

/* Definitions of variables used */
      char fltno[8];
      char str[20];
      char ch;
      int ret;
      int opt2;
      int eofcheck;

      /* Structure definition for arrivals file */
      /* This contains the flight number, scheduled */
      /* arrival time, expected arrival time, airport */
      /* from where the flight left, and remarks, which */
      /* contains current information, e.g., "landed" etc. */

      struct arrivals {
            char posn[3];
            char flight_no[8];
            char sch_arrival_time[6];
            char exp_arrival_time[6];
            char origin[15];
            char remarks[14];
      };
      /* Structure for flightcnt.dat file */
      /* which contains the current number */
      /* of flights in the arrivals file */
```

```
        struct flightcount {
        int count;
    };

    FILE *source, *target;
    FILE *fparr;
    FILE *fparr2;
    FILE *fltcnt;
    struct arrivals s1;
    struct arrivals st[17];
    struct flightcount fc;

    /* User-entered storage variables */
    char new_sch_arrival_time[6];
    char new_exp_arrival_time[6];
    char new_origin[15];
    char new_remarks[14];

    int i,opt;
    int lim;
    long int minusone = -1;
    char rollup;

    char oldname[] = "arrivals2.dat";
    char newname[] = "arrivals.dat";
    int remans;
main()
{

    /* Open flightcnt.dat file to find the current */
    /* number of flights in the file. */
    /* Store this number in opt2 */
    /* so that near to the end of the day */
    /* the boards will display fewer than 12 flights */
    fltcnt = fopen("flightcnt.dat","r");
    fread(&fc, sizeof(fc), 1, fltcnt);
```

```
printf(" Number of flights : %d", fc.count);
opt2 = fc.count;
fclose(fltcnt);

/* Open arrivals file */
fparr = fopen("arrivals.dat", "r");
printf("\n Flight\t:Sched:\tExp:\tOrigin:\t Remarks");

for (i = 0;i < 17;i++)
{
      /* Read each flight data from file sequentially */
      /* and display them */
      if(fread(&s1, sizeof(s1), 1, fparr) == 1)
      {
            /* Print flight no, sched, and expected time and flight
            origin for each flight */
            strcpy(st[i].posn,s1.posn);
            strcpy(st[i].flight_no,s1.flight_no);
            strcpy(st[i].sch_arrival_time,s1.sch_arrival_time);
            strcpy(st[i].exp_arrival_time,s1.exp_arrival_time);
            strcpy(st[i].origin,s1.origin);
            strcpy(st[i].remarks,s1.remarks);

            /* Only print the first 12 flights on the "display board" */
            if(opt2 < 12)
                  lim = opt2;
            else
                  lim = 12;
            if(i<lim)
                  printf("\n : %s\t%s\t%s\t%s\t%-12s\t%s",
                  s1.posn,s1.flight_no,s1.sch_arrival_time,
                  s1.exp_arrival_time,s1.origin,s1.remarks);
      }
}
fclose(fparr);

/* The flights can be rolled up (when a flight had landed) */
```

```
        /* or amendments made to the display */

        printf("\nroll up ? y or n \n");
        scanf("%c", &rollup);

        /*                                   */
        /* "NOT ROLL UP" SECTION OF THE PROGRAM */
        /*                                   */

        if(rollup == 'n')
        {
               notrollupfunc();
        } /* End of not rollup */

        /*                                */
        /* "ROLL UP" SECTION OF THE PROGRAM */
        /*                                */

        if(rollup == 'y')
        {
               rollupfunc();
        } /* End of rollup */

}
void notrollupfunc()
{
        /* Amendments can be made to the scheduled arrival time, */
        /* the expected arrival time, the airport of origin, */
        /* or remarks for the flight */

        /* The flight number is asked for */
        printf("\nenter flight number (max 10 ) \n");
        scanf("%s", fltno);
        printf("\n Flight number is %s",fltno);

        printf("\nenter the field you want to change \n");
        printf("\nsched = 1,Exp = 2 Origin = 3 Remarks = 4\n");
        scanf("%d", &opt);
        /* A switch command uses the number entered */
```

```
switch(opt)
{
      case 1:
            /* Change scheduled arrival time */
            printf("\nenter the new sched \n");
            scanf("%s", new_sch_arrival_time);
            printf("\nnew sched is %s",new_sch_arrival_time);

            for (i = 0;i < 17;i++)
            {
                  ret=strcmp(fltno,st[i].flight_no);
                  if(ret == 0)
                  {
                        strcpy(st[i].sch_arrival_time,new_sch_
                        arrival_time);
                        printf("\nstored sched is %s",st[i].sch_
                        arrival_time);
                        printf("\n Flight number is %s",&fltno);
                        printf("\n struct Flight number is %s",
                        st[i].flight_no);
                  }
            }
            break;
      case 2:
            /* Change expected arrival time */

            printf("\nenter the new exp \n");
            scanf("%s", new_exp_arrival_time);
            printf("\nnew exp is %s",new_exp_arrival_time);
            for (i = 0;i < 17;i++)
            {
                  ret=strcmp(fltno,st[i].flight_no);
                  if(ret == 0)
                  {
                        strcpy(st[i].exp_arrival_time,new_exp_
                        arrival_time);
```

```
                        printf("\nstored exp is %s",st[i].exp_
                        arrival_time);
                        printf("\n Flight number is %s",&fltno);
                        printf("\n struct Flight number is %s",
                        st[i].flight_no);
                }
        }
        break;
    case 3:
        /* Change airport of origin */
        printf("\nenter the new origin \n");
        scanf("%s", new_origin);
        printf("\nnew origin is %s",new_origin);
        for (i = 0;i < 17;i++)
        {
                ret=strcmp(fltno,st[i].flight_no);
                if(ret == 0)
                {
                        strcpy(st[i].origin,new_origin);
                        printf("\nstored origin is %s",st[i].origin);
                        printf("\n Flight number is %s",&fltno);
                        printf("\n struct Flight number is %s",
                        st[i].flight_no);
                }
        }
        break;
    case 4:
        /* Add remarks */

        printf("\nenter the number of the remark \n");
        printf("1 = On Approach\n");
        printf("2 = Delayed\n");
        printf("3 = Landed\n");
        scanf("%d", &remans);
        switch (remans)
        {
```

```
                case 1:
                        strcpy(new_remarks,"On Approach");
                        break;
                case 2:
                        strcpy(new_remarks,"Delayed");
                        break;
                case 3:
                        strcpy(new_remarks,"Landed");
                        break;
                default:
                        break;
        }
        printf("\n Flight number is %s",&fltno);

        printf("\nnew remarks is %s",new_remarks);
        for (i = 0;i < 17;i++)
        {
                ret=strcmp(fltno,st[i].flight_no);
                if(ret == 0)
                {
                        strcpy(st[i].remarks,new_remarks);

                        printf("\nstored remarks is %s",st[i].
                        remarks);
                        printf("\n Flight number is %s",&fltno);
                        printf("\n struct Flight number is %s",
                        st[i].flight_no);
                }
        }
        break;
    default:
        printf("\nerror \n");
} /* End of switch */

/* Output updated arrivals data */
```

```
    fparr = fopen("arrivals.dat", "w");
    if(opt2 < 12)
        lim = opt2;
    else
        lim = 12;
    printf("\n Flight\t:Sched:\tExp:\tOrigin:\t Remarks");

    for (i = 0;i < 17;i++)
    {
        strcpy(s1.posn,st[i].posn);
        strcpy(s1.flight_no,st[i].flight_no);
        strcpy(s1.sch_arrival_time,st[i].sch_arrival_time);
        strcpy(s1.exp_arrival_time,st[i].exp_arrival_time);
        strcpy(s1.origin,st[i].origin);
        strcpy(s1.remarks,st[i].remarks);
        fwrite(&s1 ,sizeof(s1),1 , fparr );

        if(i<lim)
            printf("\n : %s\t%s\t%s\t%s\t%-12s\t%s",st[i].posn,
            st[i].flight_no,st[i].sch_arrival_time,st[i].exp_arrival_
            time,st[i].origin,st[i].remarks);
    }
    fclose(fparr);
}
void rollupfunc()
{
    /* Flight to be rolled off the display, */
    /* e.g., if it has landed */
    printf("\nenter flight number  \n");
    scanf("%s", fltno);

    /* Write to a temporary file arrivals2.dat */
    /* which will then overwrite arrivals.dat */

    fparr = fopen("arrivals.dat", "r");
    fparr2 = fopen("arrivals2.dat", "w");
```

```
if(opt2 < 12)
      lim = opt2;
else
      lim = 12;

for (i = 0;i < 17;i++)
{
      eofcheck = fread(&s1, sizeof(s1), 1, fparr);
      if(eofcheck == 0)
      {
            goto exit;
      }

      if(strcmp(s1.flight_no,fltno) != 0)
      {
            fwrite(&s1 ,sizeof(s1),1 , fparr2 );
      }
      else
      {
            /* update the flight count file */

            fltcnt = fopen("flightcnt.dat","r");
            fread(&fc, sizeof(fc), 1, fltcnt);
            fclose(fltcnt);
            fc.count = fc.count-1;
            opt2 = fc.count;
            fltcnt = fopen("flightcnt.dat","w");
            fwrite(&fc, sizeof(fc), 1, fltcnt);
            fclose(fltcnt);

            fltcnt = fopen("flightcnt.dat","r");
            fread(&fc, sizeof(fc), 1, fltcnt);
            fclose(fltcnt);

      }
}
```

```
exit:
      fclose(fparr);
      fclose(fparr2);

      /* Copy arrivals2.dat to new arrivals.dat */
      remove("arrivals.dat");

      source = fopen("arrivals2.dat", "r");
      target = fopen("arrivals.dat", "w");

      while ((ch = fgetc(source)) != EOF)
            fputc(ch, target);

      fclose(source);
      fclose(target);

      /* Display updated data */

      fparr = fopen("arrivals.dat", "r");
      if(opt2 < 12)
            lim = opt2;
      else
            lim = 12;
      printf("\n Flight\t:Sched:\tExp:\tOrigin:\t Remarks");
      for (i = 0;i < lim;i++)
      {
            fread(&s1, sizeof(s1), 1, fparr);

            strcpy(st[i].posn,s1.posn);
            strcpy(st[i].flight_no,s1.flight_no);
            strcpy(st[i].sch_arrival_time,s1.sch_arrival_time);
            strcpy(st[i].exp_arrival_time,s1.exp_arrival_time);
            strcpy(st[i].origin,s1.origin);
            strcpy(st[i].remarks,s1.remarks);
```

```
            printf("\n : %s\t%s\t%s\t%s\t%-12s\t%s",st[i].posn,
            st[i].flight_no,st[i].sch_arrival_time,st[i].exp_arrival_time,
            st[i].origin,st[i].remarks);

        }
        fclose(fparr);
}
```

The following diagram shows what the arrivals display board would look like (Figure 6-1).

FLIGHT ARRIVALS

	Flight	Sched	Exp	Origin	Remarks
:1	AA1232	07:00	07:00	CHICAGO	Approaching
:2	BA123	07:05	07:05	LONDON	
:3	AA4517	07:08	07:15	BOSTON	Delayed
:4	AF123	07:10	07:10	PARIS	
:5	NH444	07:20	07:20	TOKYO	
:6	DJ144	07:22	07:22	MUMBAI	
:7	AZ2348	07:23	07:25	WASHINGTON	
:8	VS9745	07:25	07:26	TORONTO	
:9	DL5816	07:30	07:30	CHICAGO	
:10	KL5393	07:33	07:33	MANCHESTER	
:11	AZ4627	07:35	07:40	ROME	
:12	VS4677	07:40	07:40	NEW ORLEANS	

Figure 6-1. *Arrivals board example*

EXERCISES

1. Write the program, as described at the start of this chapter, to read the contents of the flightcnt.dat file and print to the user.

2. Using the following structure, write a program to create a file for airport departures.

   Create the file depcnt.dat to keep the count of departure flights.

```
struct departures {
char posn[3];
char flight_no[8];
char sch_departure_time[6];
char exp_departure_time[6];
char destination[15];
char checkingate[5];
char remarks[14];
};
```

CHAPTER 7

Power Plant Control

7.1 Simulation

This is a simulation of an industrial process. The chapter has the title "Power Plant Control," but it is just a demonstration of how any industrial process uses computers to assist in its operation.

We assume that the plant has many devices throughout its operation which monitor temperatures and flow rates at different points. In the past these would just be gauges which would have to be monitored manually at all times. So this would involve hundreds of people who would be checking these. If one of the people monitoring these was distracted and missed a high reading, it could be catastrophic. With computers, these devices can be connected to the computer and send messages to them containing values of temperatures and flow rates. These can be sent to one point where a number of people can monitor the devices on screens. The computers can be programmed to hold values for these flow rates that would be considered to be high so that when the flow rate gets close to this value an alarm can be raised for the person monitoring. When computers were first used to do this type of thing, the person monitoring would get in touch with an engineer who would investigate. Even though the computer may be able to fix the problem itself, the engineers did not feel that they could trust them. These days, people have more confidence in the reliability of computers and the software has much more control.

As with the previous two chapters, we modify how we use the program so that we can simulate what is happening. So we manually send a message to the program from the command line and we receive messages back from the program to the user. We will monitor temperatures and flow rates. The device (in this case the user) can send a message containing the current flow rate and temperature and also a high flow rate value and a high temperature value which, if reached, will cause an alarm to be raised.

P. Joyce, *Practical Numerical C Programming*, https://doi.org/10.1007/978-1-4842-6128-6_7

We will create a file which contains these values. Each device will have its own ID so the structure for each device will be as follows:

```
struct fplant {
        int ID; /* ID of device */
        float temp; /* Current temperature of device */
        float flowrate; /* Current flow rate of device */
        float hightemp; /* Maximum temperature of device */
        float highflow; /* Maximum flow rate of device */
};
```

We will define data for 17 devices in our file. The same mechanism is used as in the past two chapters. The 17 structures are preset at the start of the program and then these are written to the file using a series of `fwrite` commands. An example of the preset array structure is shown in the following:

```
struct fplant s10 = {4,10.0,23.0,50.0,50.0};
```

In this case the ID is 4, temperature is 10.0, flow rate is 23.0, high temp is 50.0, and high flow is 50.0.

Each of these structures is written to the file using an fwrite command as shown in the following for the s10 structure:

```
fwrite(&s10, sizeof(s1), 1, fp);
```

We then close the file, then reopen it, and we read the data from the file and write it to the screen for the user to check. The file is called `tempflow.bin`. The code for this is shown as follows:

```
/* createplantb.c  */
/* Industrial plant simulation */
/* Creates file */
/* Reads from file */
/* Prints out the records sequentially */

/* Power plant temperature and flow rate */

#define _CRT_SECURE_NO_WARNINGS
#include<stdio.h>
```

```
/* Structure definition for each device on file */
struct fplant {
   int ID; /* ID of device */
   float temp; /* Current temperature of device */
   float flowrate; /* Current flow rate of device */
   float hightemp; /* Maximum temperature of device */
   float highflow; /* Maximum flow rate of device */
};

int main()
{
      int i,numread;
      FILE *fp;
      struct fplant s1;
      struct fplant s2;

      /* 17 structures. One for each device */
      /* Each has preset values for each */
      /* element of the structure */

      struct fplant s10 = {4,10.0,23.0,50.0,50.0};
      struct fplant s11 = {7,11.0,34.0,51.0,50.0};
      struct fplant s12 = {9,12.0,44.0,52.0,50.0};
      struct fplant s13 = {11,13.0,25.0,53.0,50.0};
      struct fplant s14 = {14,14.0,34.0,54.0,50.0};
      struct fplant s15 = {16,15.0,51.0,55.0,50.0};
      struct fplant s16 = {17,16.0,23.0,56.0,50.0};
      struct fplant s17 = {19,17.0,44.0,57.0,50.0};
      struct fplant s18 = {23,18.0,35.0,58.0,50.0};
      struct fplant s19 = {24,19.0,40.0,59.0,50.0};
      struct fplant s20 = {27,20.0,40.0,60.0,50.0};
      struct fplant s21 = {31,21.0,42.0,61.0,50.0};
      struct fplant s22 = {32,22.0,45.0,62.0,50.0};
      struct fplant s23 = {35,23.0,47.0,63.0,50.0};
      struct fplant s24 = {38,24.0,41.0,63.0,50.0};

      struct fplant s28 = {44,28.0,54.0,68.0,50.0};
      struct fplant s29 = {47,29.0,58.0,69.0,50.0};
```

```
/* Open the file */

fp = fopen("tempflow.bin", "w");

/* Write details of each ID to file*/
/* from the structures defined earlier */

fwrite(&s10, sizeof(s1), 1, fp);
fwrite(&s11, sizeof(s1), 1, fp);
fwrite(&s12, sizeof(s1), 1, fp);
fwrite(&s13, sizeof(s1), 1, fp);
fwrite(&s14, sizeof(s1), 1, fp);
fwrite(&s15, sizeof(s1), 1, fp);
fwrite(&s16, sizeof(s1), 1, fp);
fwrite(&s17, sizeof(s1), 1, fp);
fwrite(&s18, sizeof(s1), 1, fp);
fwrite(&s19, sizeof(s1), 1, fp);
fwrite(&s20, sizeof(s1), 1, fp);
fwrite(&s21, sizeof(s1), 1, fp);
fwrite(&s22, sizeof(s1), 1, fp);
fwrite(&s23, sizeof(s1), 1, fp);
fwrite(&s24, sizeof(s1), 1, fp);

fwrite(&s28, sizeof(s1), 1, fp);
fwrite(&s29, sizeof(s1), 1, fp);

/* Close the file */

fclose(fp);

/* Reopen the file */

fp=fopen("tempflow.bin", "r");

/* Read and print out all of the records on the file */

for(i=0;i<17;i++)
{
        numread=fread(&s2, sizeof(s2), 1, fp);
```

```
            if(numread == 1)
            {

                    printf("\nID : %d temp : %f flow rate : %f high temp :
                    %f high flow : %f", s2.ID,s2.temp,s2.flowrate,s2.
                    hightemp,s2.highflow);

            }
            else {
                    /* If an error occurred on read, then print out message */

                            if (feof(fp))

                            printf("Error reading tempflow.bin : unexpected end
                            of file fp is %p\n",fp);

                            else if (ferror(fp))
                    {
                            perror("Error reading tempflow.bin");
                    }
            }

        }
        /* Close the file */

        fclose(fp);

}
```

7.2 Monitoring Safety Levels

We now want to write a program to enter values for a specific device and test that the value entered is not outside the allowed limit. If it is outside, then we need to print an alert message.

We first read the file and print out data held for each device ID. Then we need to ask the user to enter the ID of the device and the current temperature. In the program, there is a "do" loop where the ID is entered. If the user enters an ID which is not on the file, then an error message is sent. When the user has entered a valid ID, the code exits the "do" loop. We test the current temperature against the limit held in the record. If it is outside this limit, we output an alert. We then ask the user to enter the flow rate for the same device ID and we test this against the flow rate limit.

The code for this is shown as follows:

```
/* plantb.c */
/* Industrial plant simulation */
/* Finds specific records and prints them */
/* Checking power plant temperature and flow rate */
/* against acceptable levels */
#define _CRT_SECURE_NO_WARNINGS
#include <stdio.h>
#include <math.h>
#include <stdlib.h>
#include <time.h>

/* Structure definition for each device on file */
struct fplant {
   int ID; /* ID of device */
   float temp; /* Current temperature of device */
   float flowrate; /* Current flow rate of device */
   float hightemp; /* Maximum temperature of device */
   float highflow; /* Maximum flow rate of device */
};

void main()
{
      FILE *fp;

      struct fplant s2;
      struct fplant st[17];
      int i;
      int currentID;
      int IDfound;
      float currenttemp,currentflow;
      /* Open tempflow.bin file */
```

```
fp = fopen("tempflow.bin", "r");
for (i = 0;i < 17;i++)
{
      /* Read each pressure data from file sequentially */
      fread(&s2, sizeof(s2), 1, fp);
      /* Print pressure data each component */

      st[i].ID = s2.ID;
      st[i].temp = s2.temp;
      st[i].flowrate = s2.flowrate;
      st[i].hightemp = s2.hightemp;
      st[i].highflow = s2.highflow;

      printf("\nID : %2d temp : %f flow rate : %f high temp : %f high
      flow : %f", s2.ID,s2.temp,s2.flowrate,s2.hightemp,s2.highflow);
}

fclose(fp);

/* User asked to enter the ID being monitored */
/* Go round "do loop" until a valid ID is entered */
IDfound=0;
do {

      /* Ask user to enter ID */
      printf("\nenter ID \n");
      scanf("%d", &currentID);

      printf("\n ID is %d",currentID);
      for (i = 0;i < 17;i++)
      {

            if(currentID == st[i].ID)

            {
                  /* Valid ID found */
                  IDfound=1;
                  break;
            }

      }
```

```
        if(IDfound==0)
            printf("\nID not found");

    } while( IDfound==0);

    /* User asked to enter the current temperature being monitored */

    printf("\nenter current temperature \n");

    scanf("%f", &currenttemp);

    printf("\n current temperature is %f",currenttemp);

    /* Current temperature checked against range of temperature */
    /* An Alert is displayed if the temperature is outside the range */

        for (i = 0;i < 17;i++)
        {

            if(currentID == st[i].ID)

            {

                printf("\n high temp is %f",st[i].hightemp);

                if(currenttemp > st[i].hightemp)
                    printf("\n ALERT! Temperature is above upper
                    limit");
            }

        }

    /* User asked to enter the flow rate being monitored */

    printf("\nenter current flow rate \n");
    scanf("%f", &currentflow);

    printf("\n current flow rate is %f",currentflow);

    /* Flow rate checked against limits */
    /* An alert is displayed if the flow rate is outside the range */
```

```
            for (i = 0;i < 17;i++)
            {

                    if(currentID == st[i].ID)

                    {

                            printf("\n high flow rate is %f",st[i].highflow);

                            if(currentflow > st[i].highflow)
                                    printf("\n ALERT! Flow rate is above upper
                                    limit");

                    }

            }

}
```

Our final program will update our `tempflow.bin` file.

We open the file and print out all of the records in the file. We then ask the user to enter the ID of the device being amended. We then find the ID from the file. Then we ask the user to enter the new temperature. At this point, the file pointer is pointing to the next record in the file, so we need to back it up to point to the current record which is the one we want to amend. We do this using the `fseek(fp,minusone*sizeof(s2),SEEK_CUR);` command. Then we can write the new temperature for this record and close the file.

To explain this mechanism of backup, look at the following situation.

We are reading through the file and we have just read the second entry in the file, so the file pointer is now pointing to the third entry as shown in the following:

```
4       10.0     23.0     50.0
7       11.0     34.0     51.0
9       12.0     44.0     52.0     ← file pointer
11      13.0     25.0     53.0
14      14.0     34.0     54.0
```

The fread function `fread(&s2, sizeof(s2), 1, fp);` will have placed the details for the second structure (ID = 7) into the structure variable s2. If we now want to update the current flow rate of ID = 7 from 34.0 to 47.0 in s2 and we want to write this back to the file, we need to move the pointer back to pointing at the ID = 7 structure of the file. We do this using `fseek(fp,minusone*sizeof(s2),SEEK_CUR);`.

The `minusone*sizeof(s2)` part of the instruction tells the file pointer to move back by the length of the s2 structure. After the instruction, the pointer will be pointing at the ID=7 file entry as shown in the following:

```
4     10.0    23.0    50.0
7     11.0    34.0    51.0     ← file pointer
9     12.0    44.0    52.0
11    13.0    25.0    53.0
14    14.0    34.0    54.0
```

Now we can write the new data to the file using `fwrite(&s2, sizeof(s2), 1, fp);`. The fwrite instruction writes the new contents of s2 to the position in the file pointed to. The new data is then printed to the user. This completes the updating mechanism of the program and the flow rate of ID = 7 is updated to 47.0 as shown. After the `fwrite`, the pointer will be pointing to the next item.

```
4     10.0    23.0    50.0
7     11.0    47.0    51.0
9     12.0    44.0    52.0     ← file pointer
10    13.0    25.0    53.0
14    14.0    34.0    54.0
```

The code for this is shown as follows:

```
/* plantbam.c */
/* Industrial plant simulation */
/* Power plant temperature and flow rate */
/* Allows amendments to tempflow.bin file */
/* Tests if the amendment is above the hightemp */
/* value and outputs an alert if it is */
#define _CRT_SECURE_NO_WARNINGS
#include <stdio.h>
#include <math.h>
#include <stdlib.h>
#include <time.h>
```

```
/* Structure definition for each device on file */
struct fplant {
   int ID; /* ID of device */
   float temp; /* Current temperature of device */
   float flowrate; /* Current flow rate of device */
   float hightemp; /* Maximum temperature of device */
   float highflow; /* Maximum flow rate of device */
};

int main()
{
      FILE *fp;

      struct fplant s2;

      int i;

      int IDtoamend; /* User-entered ID variable */
      float fnewtemp;
      long int minusone = -1;
      int IDfound;
      /* Open tempflow.bin file */

      fp = fopen("tempflow.bin", "r");
      for (i = 0;i < 17;i++)
      {
            /* Read each pressure data from file sequentially */
            fread(&s2, sizeof(s2), 1, fp);
            /* Print pressure data each component */

            printf("\nID : %2d temp : %f flow rate : %f high temp : %f high
            flow : %f", s2.ID,s2.temp,s2.flowrate,s2.hightemp,s2.highflow);
      }

      fclose(fp);

      /* User asked to enter the ID being monitored */
      /* Go round "do loop" until a valid ID is entered */
      IDfound=0;
```

```
do {
      fp = fopen("tempflow.bin", "r+");
      /* Ask user to enter ID */
      printf("\nenter ID \n");
      scanf("%d", &IDtoamend);

      printf("\n ID is %d",IDtoamend);
      for (i = 0;i < 17;i++)
      {
            fread(&s2, sizeof(s2), 1, fp);
            if(IDtoamend == s2.ID)

            {
                  /* Valid ID found */
                  IDfound=1;
                  break;
            }

      }

      if(IDfound==0)
            printf("\nID not found");

      fclose(fp);
} while( IDfound==0);

fp = fopen("tempflow.bin", "r+");
/* Loop of 17 items in tempflow.bin file */
/* Need to find the user-entered ID */
for (i = 0;i < 17;i++)
{
      fread(&s2, sizeof(s2), 1, fp);
      if(IDtoamend == s2.ID)

      {
            /* Correct ID found in file */
            /* User asked to enter the new temperature being
            monitored */
```

```
                printf("\nenter new temperature   \n");
                scanf("%f", &fnewtemp);

                /* Print out confirmation of temperature to user */
                printf("\n new temperature is %f",fnewtemp);

                /* Store new temperature in file */
                s2.temp = fnewtemp;

                /* File updated with new temperature */
                /* As file pointer is currently pointing */
                /* to the next record in the file, we must */
                /* go back by 1 (minusone) to update the */
                /* correct record */
                fseek(fp,minusone*sizeof(s2),SEEK_CUR);
                fwrite(&s2, sizeof(s2), 1, fp);

                /* Print out the new values for the device */
                printf("\nID : %d temp : %f flow rate : %f high temp :
                %f high flow : %f", s2.ID,s2.temp,s2.flowrate,s2.
                hightemp,s2.highflow);

                break;

            }

        }

        fclose(fp);

}
```

You may have realized that after making the amendment to the `tempflow.bin` file in the preceding program, the next thing we should do is check that the new temperature is within the limits. So at this point, we would call the previous program to do this. A more realistic alternative would be to have the two programs combined so that you checked the new level immediately after updating. This is given as an exercise.

EXERCISES

1. Create a file to show the pressures in devices. Your program will be similar to the `createplantb.c` program in this chapter. Your structure defining each record in the file should look like this:

```
struct fpress {
int ID; /* ID for the device */
float llimit; /* Lower limit for pressure */
float press; /* Current pressure */
float ulimit;  /* Upper limit for pressure */
};
```

   Then you need to preset your structures similar to the following:

```
struct fpress s10 = {4, 10.0, 23.0, 50.0};
```

   Write these structures to the file. Then close the file, reopen it, and then read and print out all of the records in the file.

2. Write a program similar to `plantb.c` to read your file from question 1 and then ask the user to enter the device ID and its current pressure. Test this entered pressure and compare it to the allowed upper and lower limits. Output an error message if the pressure is outside either of these limits.

3. Write a program to amend the temperature, as in the program in the chapter, but then test the temperature against the hightemp value and output an alert if it is above this value.

PART III

Physics Applications

CHAPTER 8

Energy Transfer

8.1 Potential and Kinetic Energy Simulation

This is a piece of simple physics. We will look at the energy of a heavy ball which is dropped from a 10 meter tower. The ball is relatively heavy (10kg), and it is suspended from a cable at the top of the tower. When the cable is cut, the ball will accelerate under the force of gravity to the ground. We can make some calculations of the energy the ball has.

We will use some of the basic equations of motion.

$$s = u*t + 0.5*f*t^2$$

$$v = u + f*t$$

where

s is the distance traveled

u is the initial velocity

t is the time

f is the acceleration

v is the final velocity

So in our first equation if we know the initial velocity, the time, and the acceleration, we can calculate the distance traveled.

In the second equation if we know the initial velocity, the acceleration, and the time, we can calculate the final velocity.

We also use two energy equations:

$$\text{Potential energy} = m*g*h$$

$$\text{Kinetic Energy} = 0.5*m*v^2$$

P. Joyce, *Practical Numerical C Programming*, https://doi.org/10.1007/978-1-4842-6128-6_8

where

g is the acceleration of gravity

h is the height above ground

m is the mass of the object

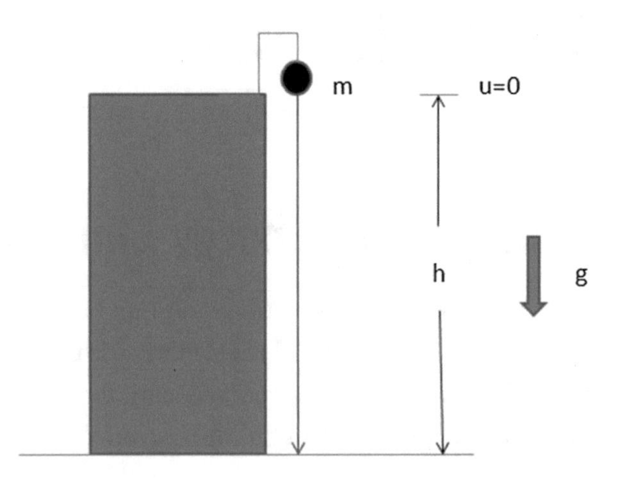

Figure 8-1. *Ball falling from tower*

Figure 8-1 is a diagram of the ball at the top of the tower. The acceleration in this case is the acceleration due to gravity and is normally represented by g.

We know that the initial velocity of the ball (just as the cable is cut) is zero. We assume that we can ignore air resistance. There will be some air resistance but it will be small compared with the other factors. We can make our calculations every 0.1 seconds. So from our formula

$$v = u + f*t$$

we have u=0, f is the acceleration of gravity, g, which is 9.8 m/s^2, and t which is 0.1. If we substitute these values into the equation, we get v, the velocity after 0.1 seconds,

$$\text{so } v = 0.98 \text{ m/s}$$

Now we can use our other formula

$$s = u*t + 0.5*f*t^2$$

where s is the height traveled in 0.1 seconds, so for u = 0, f = 9.8 m/s^2, and t = 0.1, we get s = 0.049m.

The ball will have traveled 0.049m in the first 0.1 seconds.

So its new height above the ground will be the original height –0.049.

8.2 Convert Theory to Code

In our program we can set up a loop for which we increment the time by 0.1 seconds and we find the new velocity for each iteration of the loop. We also find the height fallen so that we can calculate the new height above the ground after each iteration. We can use these two values in our energy formulas

$$PE = m*g*h$$

$$KE = 0.5*m*v^2$$

so that after each iteration of the loop, we can store the PE and KE in a file.

The code for this is shown as follows:

```
/* peke.c */
/* Potential energy vs. kinetic energy */
/* */
/* */
#define _CRT_SECURE_NO_WARNINGS
#include <stdio.h>
#include <math.h>
#include <stdlib.h>

void main()
{

      int i;
      double m,g,t,h,hn,KE,PE;
      double u,v;

      FILE *fptr;

      fptr=fopen("peke.dat","w");

      /* Initialize the variables from the formulas */

      m=10.0; /* Preset mass (kg) value */
      g=9.8; /* Preset acceleration of gravity (m/s²) value */
      h=10.0; /* Preset height (m) value */
      t=0.1; /* Preset time division (s) value */
      u=0.0; /* Preset initial velocity (m/s) value */
```

```
    for(i=0;i<100;i++)
    {
        v=u+g*t; /* Find velocity v from initial velocity, accel. of
                gravity, and time */
        KE=0.5*m*pow(v,2); /* Find kinetic energy from mass and
                        velocity */

        hn=u*t+0.5*g*pow(t,2); /* Find new height after time t */

        h=h-hn; /* New height after falling hn meters */

        PE=m*g*h; /* Find potential energy */

        u=v; /* Set the initial velocity for the next increment of the
            loop to the current velocity */

        /* If h = 0.0, then we have reached the ground */
        if(h<=0.0)
            break;
        fprintf(fptr,"%lf\t%lf\n",KE,PE);

    }

    fclose(fptr);
}
```

The potential energy and kinetic energy values for each iteration of the loop are stored in the file peke.dat.

The output from this is shown as follows in Figure 8-2.

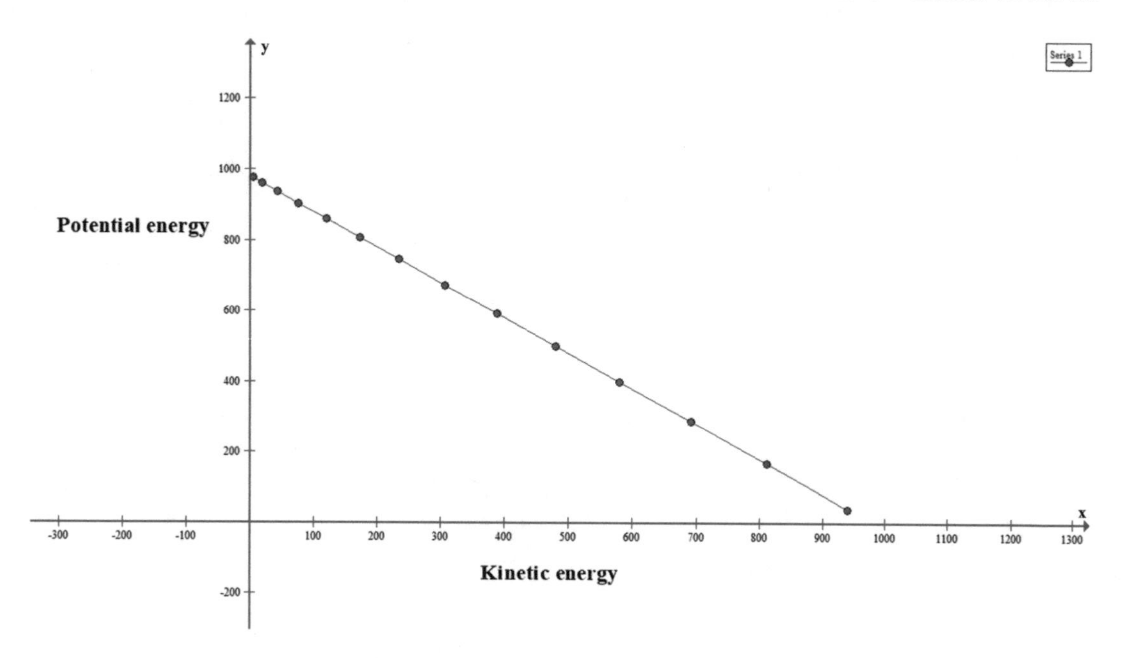

Figure 8-2. *Potential and kinetic energy*

We can see that the potential energy decreases at the same rate as the kinetic energy increases. At the start the potential energy is high and the kinetic energy is zero. At the end the potential energy is zero and the kinetic energy is large. So we can say that energy is conserved. Strictly speaking, some of the energy is transferred to the air, but this is small so we will ignore it in this case. Neil Armstrong and Buzz Aldrin could have done this experiment on the Moon where air resistance would be zero.

We can make a small modification to the program so that we add the kinetic energy to the potential energy after each iteration of the loop. We can write these values to a separate file and print these on the same graph as earlier. This is given as an exercise.

EXERCISES

1. Amend your energy program to keep a total of kinetic energy + potential energy after each iteration. Write these to a different file. Print the output from the two files on the same graph.

CHAPTER 9

Pendulum Simulation

9.1 Pendulum Theory

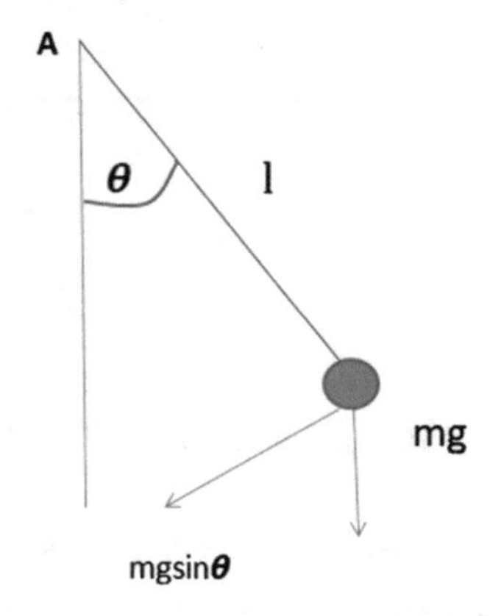

Figure 9-1. *Pendulum*

The preceding diagram shows a pendulum of length l suspended from a point A. The pendulum has a mass m attached to it and it is held at an angle θ to the vertical.

We want to find out how this angle and the angular velocity of the pendulum vary with time. If you imagine a pendulum as in Figure 9-1, then when it is released the mass moves toward the left as shown. The angle, θ, gets smaller as the mass moves to its lowest point. When it reaches its lowest point, the angle is zero. The mass continues until it reaches a point to the left of the diagram – a mirror image of the diagram. At this point, the angle is θ, but on the other side, so we say that it is - θ. It then moves back to its original position. The following graph, Figure 9-2, represents this movement. It shows the variation of theta with time.

The point, **a**, is the start point. Here, we say theta is **1**.

Point **b** is where the mass is at the bottom of its trajectory when theta is **0**.

At **c**, theta is **-1**. This is where the pendulum is to the left of the diagram.

P. Joyce, *Practical Numerical C Programming*, https://doi.org/10.1007/978-1-4842-6128-6_9

At **d,** theta is **0** again. The mass is at the bottom of its trajectory.

At **e,** the mass is back to its starting position.

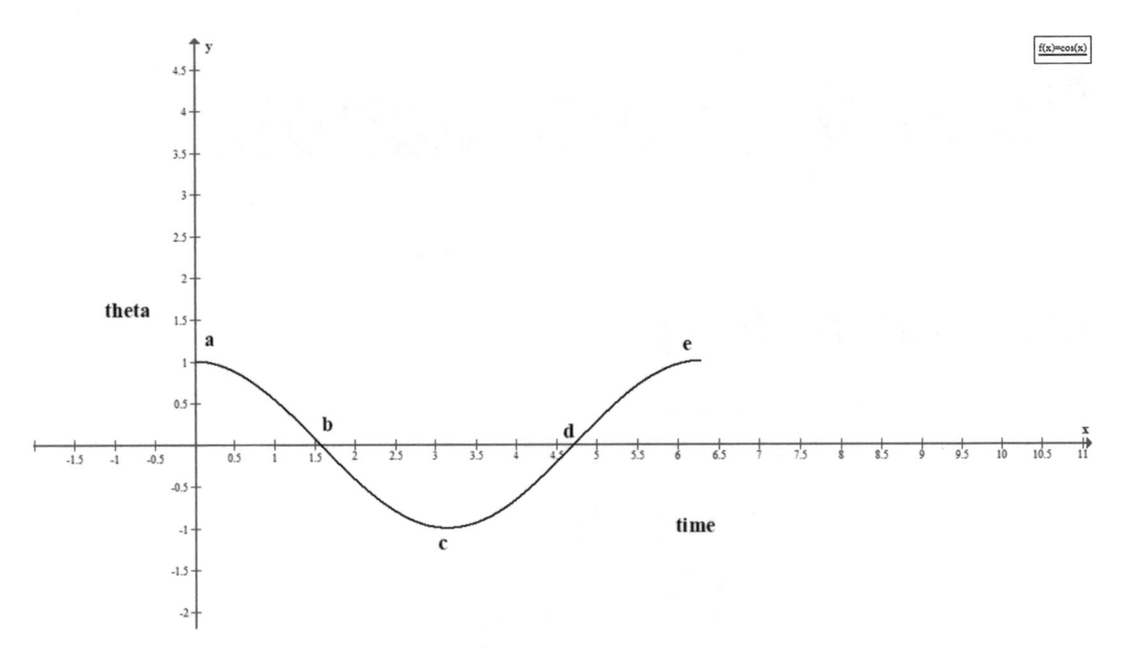

Figure 9-2. *Pendulum angle (theta) variation with time*

So the variation of $\boldsymbol{\theta}$ with time can be represented as a cosine curve as shown.

When the pendulum is released, it moves in the direction shown with a force of mgsin $\boldsymbol{\theta}$.

As force can be defined as mass multiplied by acceleration and as acceleration is the second derivative of distance with respect to time, we can write

$$\text{Force} = m^*(d^2s / dt^2)$$

But as, in this case, as force is mgsin$\boldsymbol{\theta}$, then we can combine these equations and write

$$m^*(d^2s / dt^2) = mg\sin\boldsymbol{\theta}$$

$$\text{or } (d^2s / dt^2) = g\sin\boldsymbol{\theta}$$

$$\text{or for small } \boldsymbol{\theta} \quad (d^2s / dt^2) = g\boldsymbol{\theta} \qquad (1)$$

But in the case of our pendulum diagram earlier, the distance the mass moves in the direction shown is given by

$$s = l\boldsymbol{\theta}$$

Differentiating this twice with respect to t, we get

$$(d^2s/dt^2) = l^*(d^2\boldsymbol{\theta} / dt^2)$$

Combining this with our equation (1) earlier, we get

$$l^*(d^2\boldsymbol{\theta} / dt^2) = g\boldsymbol{\theta}$$

or $$(d^2\boldsymbol{\theta} / dt^2) = (g/l)^* \boldsymbol{\theta} \qquad (2)$$

$$\text{but } (d\, \boldsymbol{\theta} / dt) = \boldsymbol{\omega} \text{ (angular velocity)}$$

$$\text{So } (d^2\boldsymbol{\theta} / dt^2) =(d\, \boldsymbol{\omega} / dt)$$

So we can write equation (2) as

$$(d\, \boldsymbol{\omega} / dt) = (g / l\,)^* \boldsymbol{\theta}$$

Now, we have two differential equations

$$(d\, \boldsymbol{\theta} / dt) = \boldsymbol{\omega} \qquad (3)$$

$$\text{and } (d\, \boldsymbol{\omega} / dt) = (g / l\,)^* \boldsymbol{\theta} \qquad (4)$$

The first equation relates the change in the angle, $\boldsymbol{\theta}$, with time. The second equation relates the change of the angular velocity, $\boldsymbol{\omega}$, with time. These are the two relationships we want to investigate. We can solve these by the Euler method.

9.2 Euler Method

The Euler method relates a function to its derivatives. The relationship between a function and its first derivative is shown in Figure 9-3.

Here, the curved line is our function and the slanted line (ab), which just touches it, is the curves gradient at the point where it touches the curve. This gradient is the first derivative evaluated at that point.

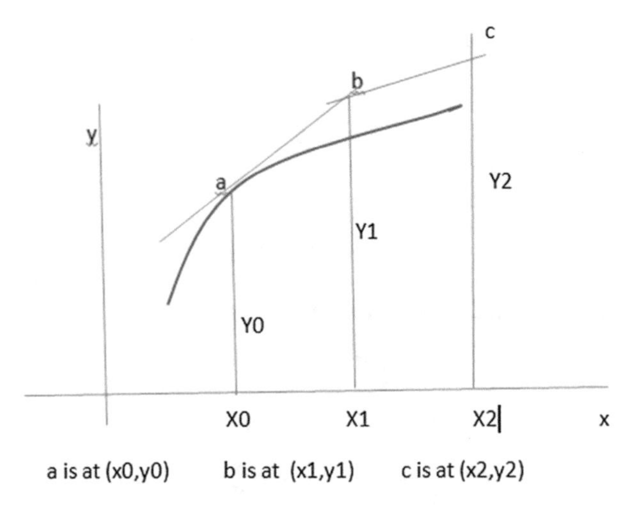

Figure 9-3. *Euler method*

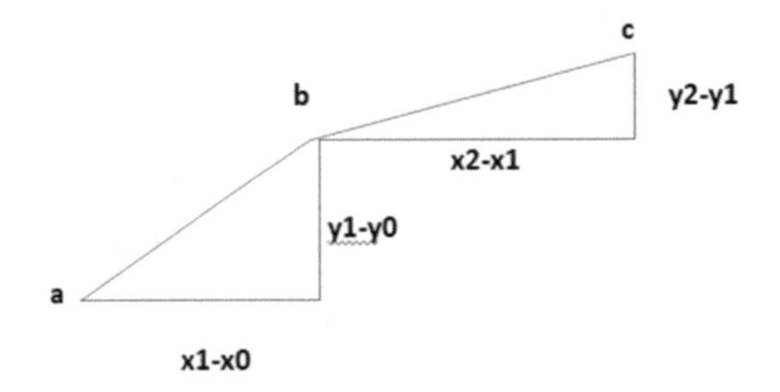

Figure 9-4. *Euler method analysis*

The preceding diagram, Figure 9-4, shows the lines ab and bc. We have projected horizontal lines from a and b and vertical lines from b and c to produce two triangles. For the left-hand triangle, we can see from the graph on the last page that the length of the base must be x1-x0 and the length of the perpendicular line from the base is y1-y0. Doing a similar thing on the right-hand triangle gives us a base of x2-x1 and a perpendicular of y2-y1.

As the gradient of each of these triangles is just the length of the perpendicular side divided by the length of the base, we can say that

Gradient of left triangle = (y1-y0)/(x1-x0)

Gradient of right triangle = (y2-y1)/(x2-x1)

The gradient of the triangles at each point is just the derivative evaluated at that point, which we can write as

$$(f(x) - f(a)) / (x-a) = f'(a)$$

where f(x) is the y value of the function at **x**

f(a) is the y value of the function at **a**

(x-a) is the length of the base of the triangle

f′(a) is the derivative or gradient function

If we have the same length of the base for each triangle, we can call it **h**. So we can rewrite (f(x) - f(a)) / (x-a) = f′(a)

as

$$(f(x)- f(a))/h = f'(a)$$

or

$$f(x)-f(a) = hf'(a)$$

or

$$\mathbf{f(x) = f(a) + hf'(a)}$$

What we can do with this formula is set initial values of f(x) and **a** and set up a loop in which we increase the **a** value by **h** for each pass of the loop.

So as we increment our x value, we calculate a new f(x) or y value. So we could write the preceding equation as

$$\mathbf{y_{n+1} = y_n + hf'(x)}$$

This is the Euler method of finding f(x), if we already know f′(x) and initial values of f(x) and a.

We can solve our two pendulum differential equations, (3) and (4), using the Euler method.

We just replace $y_{n+1} = y_n + hf'(x)$ by

$$\boldsymbol{\omega}_{n+1} = \boldsymbol{\omega}_n + (-g / l)^* \boldsymbol{\theta}^* dt \qquad (5)$$

and

$$\boldsymbol{\theta}_{n+1} = \boldsymbol{\theta}_n + \boldsymbol{\omega}_n^* dt \qquad (6)$$

So we can set initial values for $\boldsymbol{\theta}_n$ and $\boldsymbol{\omega}_n$ and use the preceding formulas, (5) and (6), in a loop in our program with increments of time dt in the loop.

```
/* pendme.c */
/*
 Simple Euler method
 */
```

```
#define _CRT_SECURE_NO_WARNINGS
#include <math.h>
#include <stdio.h>

void main()
{
      FILE *fptr;
      FILE *fptr2;
      int i,npoints;

      double length,g,dt,omega[250],theta[250],time[250];

      /* Two output files - one measuring omega, one measuring theta */

      fptr=fopen("pendout.dat","w");
      fptr2=fopen("pendoutb.dat","w");

      /* Preset the parameters from the formula */

      length=1.0; /* Preset length of pendulum (l) */
      g=9.8; /* Preset acceleration of gravity (m/s^2) */
      npoints=250; /* Preset number of points in loop */
      dt=0.04; /* Preset time interval (s) */

      /* Clear storage arrays to zero */
      for(i=0;i<npoints;i++)
      {
            omega[i]=0.0;
            theta[i]=0.0;
            time[i]=0.0;
      }

      /* Preset initial theta and omega values */
      theta[0]=0.2;
      omega[0]=0.0;

      /* Euler method */
      /* ωn+1  =  ωn + (-g / l )* θ*dt */
      /* and */
      /* θn+1 =  θn + ωn*dt */

      for(i=0;i<npoints;i++)
```

```
    {
            omega[i+1]=omega[i]-(g/length)*theta[i]*dt;
            theta[i+1]=theta[i]+omega[i]*dt;
            time[i+1]=time[i]+dt;

            fprintf(fptr,"%lf\t%lf\n",time[i+1],theta[i+1]);
            fprintf(fptr2,"%lf\t%lf\n",time[i+1],omega[i+1]);
    }
    fclose(fptr);
    fclose(fptr2);
}
```

The output from the two files is shown in Figure 9-5. From our program, the pendout.dat file monitors the theta values and pendoutb.dat monitors the omega values.

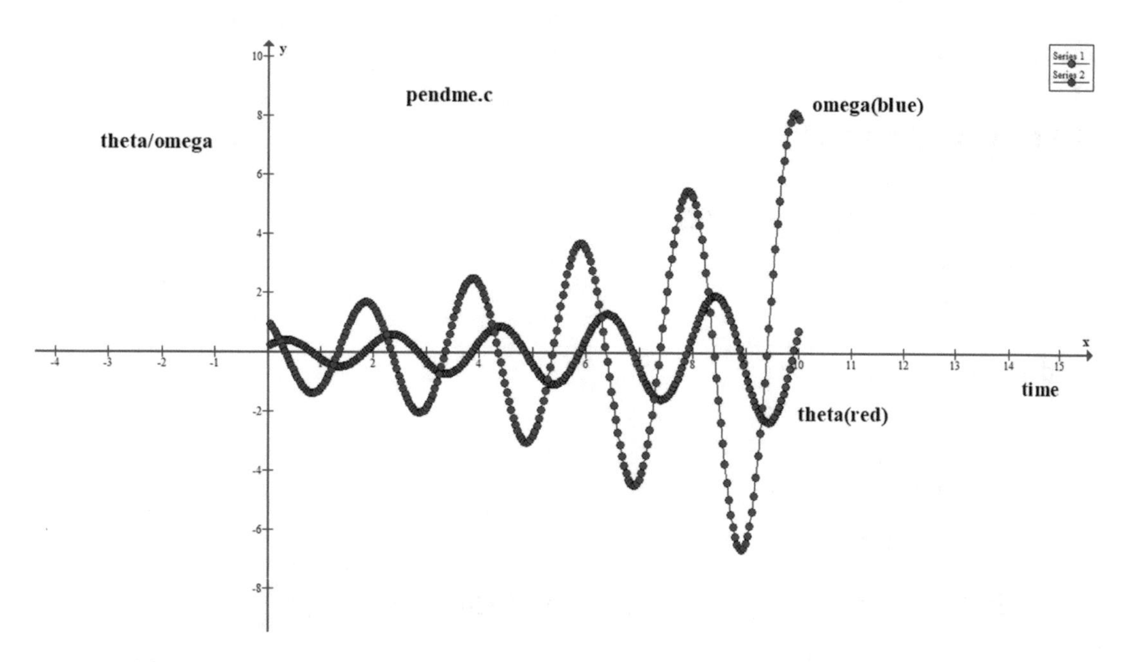

Figure 9-5. *Euler method output*

These two graphs look a bit strange. They both show the correct cyclic movement of the pendulum, but the amplitude of each is increasing. This would mean that the swings of the pendulum would increase on every cycle – an increase in energy with no external energy input. If this was correct, it could solve all of the world's energy problems. However, if you do a simple experiment of your own with a pendulum, you will see that the amplitude gradually sets smaller and eventually the pendulum will stop.

9.3 Euler-Cromer Method

So the conclusion is that the Euler method of simulation is not correct as far as the amplitude is concerned. A small change to the Euler method, called the Euler-Cromer method, corrects this problem.

$$\boldsymbol{\omega}_{n+1} = \boldsymbol{\omega}_n + (-g / l)* \boldsymbol{\theta}*dt \quad (7)$$

$$\boldsymbol{\theta}_{n+1} = \boldsymbol{\theta}_n + \boldsymbol{\omega}_{n+1}*dt \quad (8)$$

If you compare equations (7) and (8) with equations (5) and (6), you will see that equation (7) is exactly the same as equation (5), and the only difference between equation (8) and equation (6) is that we have replaced the $\boldsymbol{\omega}_n$ in equation (6) with $\boldsymbol{\omega}_{n+1}$ in equation (8).

So we modify our program as shown in the following:

```
/* pendme2.c */
/*
 Euler-Cromer method
 */
#define _CRT_SECURE_NO_WARNINGS
#include <math.h>
#include <stdio.h>

void main()
{
      FILE *fptr;
      FILE *fptr2;
      int i,npoints;

      double length,g,dt,omega[250],theta[250],time[250];

      /* Two output files - one measuring omega, one measuring theta */

      fptr=fopen("pendout2.dat","w");
      fptr2=fopen("pendout2b.dat","w");

      /* Preset the parameters from the formula */

      length=1.0; /* Preset length of pendulum (l) */
      g=9.8; /* Preset acceleration of gravity (m/s2) */
```

```
        npoints=250; /* Preset number of points in loop */
        dt=0.04; /* Preset time interval (s) */

        /* Clear storage arrays to zero */
        for(i=0;i<npoints;i++)
        {
               omega[i]=0.0;
               theta[i]=0.0;
               time[i]=0.0;
        }

        /* Preset initial theta and omega values */
        theta[0]=0.2;
        omega[0]=0.0;

        /* Euler-Cromer method */
        /* ωn+1 =  ωn + (-g / l )* θ*dt */
        /* and */
        /* θn+1 = θn + ωn+1*dt*/

        for(i=0;i<npoints;i++)
        {
               omega[i+1]=omega[i]-(g/length)*theta[i]*dt;
               theta[i+1]=theta[i]+omega[i+1]*dt;
               time[i+1]=time[i]+dt;

               fprintf(fptr,"%lf\t%lf\n",time[i+1],theta[i+1]);
               fprintf(fptr2,"%lf\t%lf\n",time[i+1],omega[i+1]);
        }
        fclose(fptr);
        fclose(fptr2);
}
```

The output from this program is shown in the following, Figure 9-6 and Figure 9-7. We still have the correct sine and cosine curves for the angular velocity and angular displacement as we had before, but now we get a constant amplitude. If we are ignoring air resistance to the pendulum and friction at the point of contact of the pendulum cord at point A in our diagram, then this is correct.

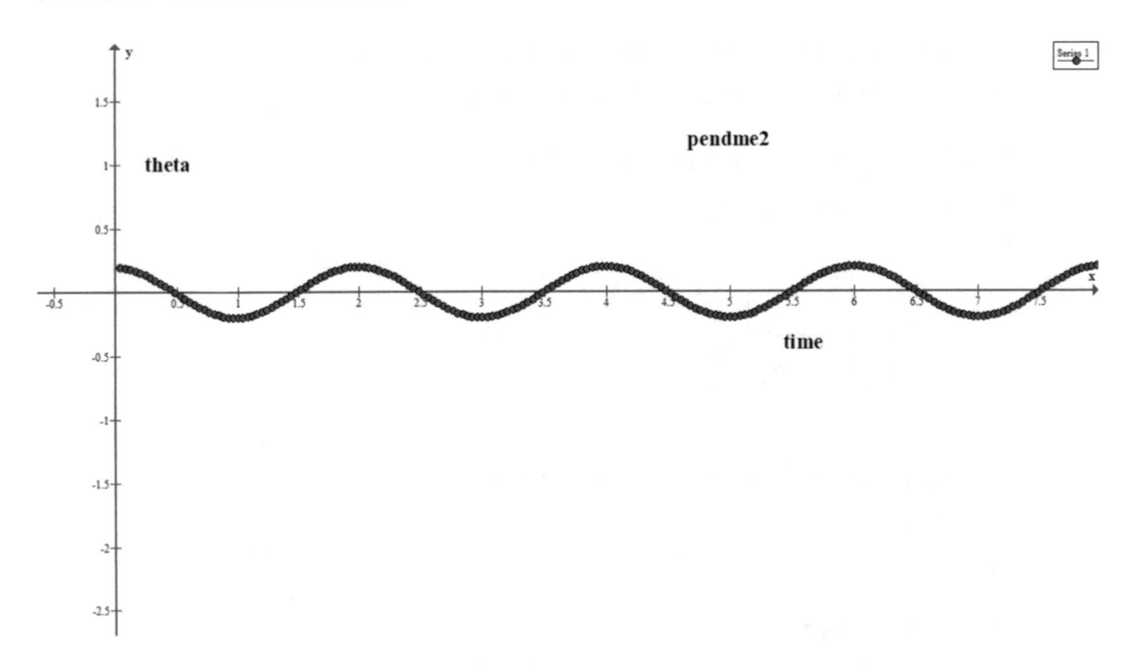

Figure 9-6. *Euler-Cromer output for theta*

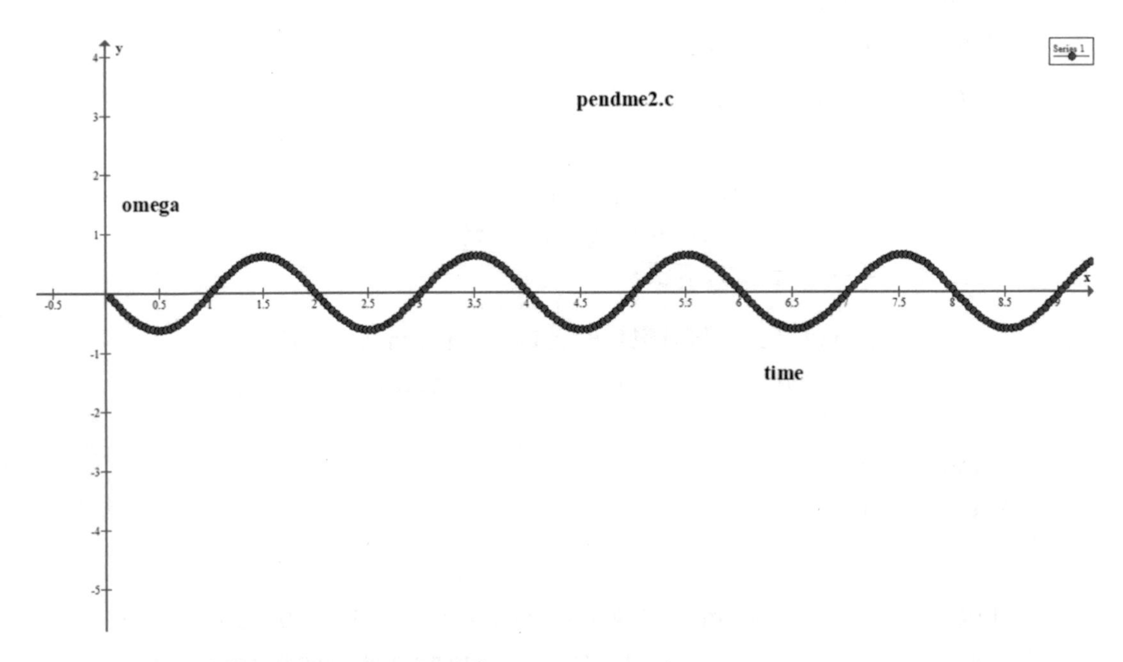

Figure 9-7. *Euler-Cromer output for omega*

EXERCISES 9

1. Amend your Euler-Cromer program to change the length of the cord to be 2m instead of 1m. Change the names of the two output files from the original program. Run your program and then print the output file for omega values. One the same graph, print out the omega value graph from the original Euler-Cromer program.

   Compare the two graphs. What has been the effect of doubling the length of the cord?

CHAPTER 10

Center of Mass

10.1 Center of Mass Theory

The Center of Mass of a body is the place where it can be said that all of the mass of the body seems to act. We can do a simple experiment to demonstrate this with a normal dinner plate. The following, Figure 10-1, is a diagram of our dinner plate. The two lines on the plate meet at its center. If we pick up the plate and carefully place our forefinger at the center point and then slowly move the plate with our free hand until the plate is horizontal, then we should be able to balance the plate on our finger and let go of the plate with our free hand.

The center of the plate is the center of mass.

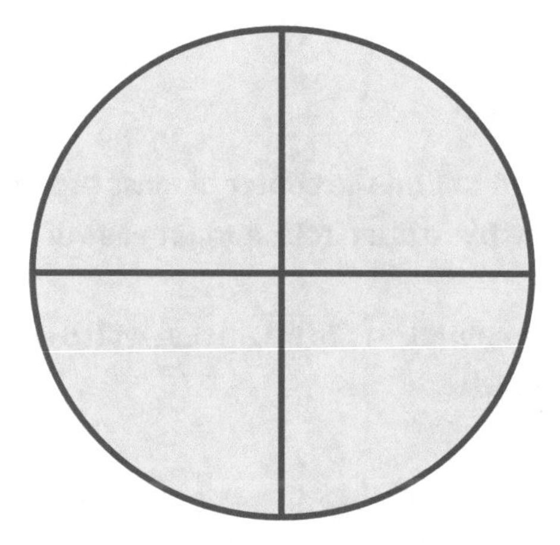

Figure 10-1. *Circular plate*

The following diagram is an oval-shaped dinner plate. It also has a horizontal line and a vertical line meeting at the center. We should be able to repeat our experiment with the oval plate and it should balance on one finger.

P. Joyce, *Practical Numerical C Programming*, https://doi.org/10.1007/978-1-4842-6128-6_10

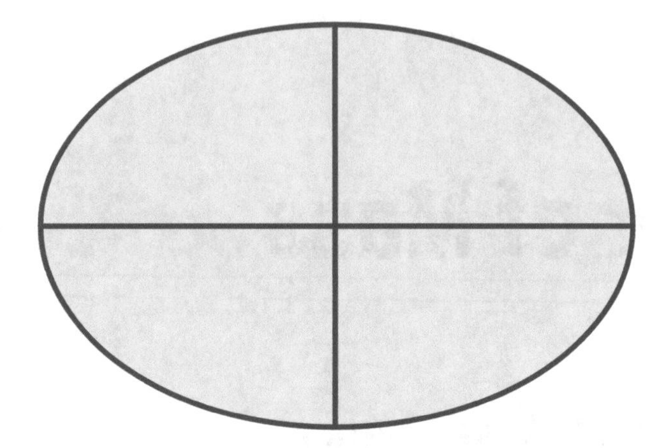

Figure 10-2. *Oval plate*

We could do the same experiment with a meter rule, shown in Figure 10-3, as long as we can find its center of mass. This shouldn't be too difficult, but in physics and engineering sometimes we need to find the center of mass of nonsymmetrical objects.

Figure 10-3. *Meter rule*

10.2 Circular Plate

In our first example, we want to find the center of mass of a circular plate as earlier. We will just assume that the plate is a circle, and for ease of demonstration, we will do this in 2D.

We will have the circle centered on the origin and with a radius of 2 units. We will write our output to file cofmc.dat.

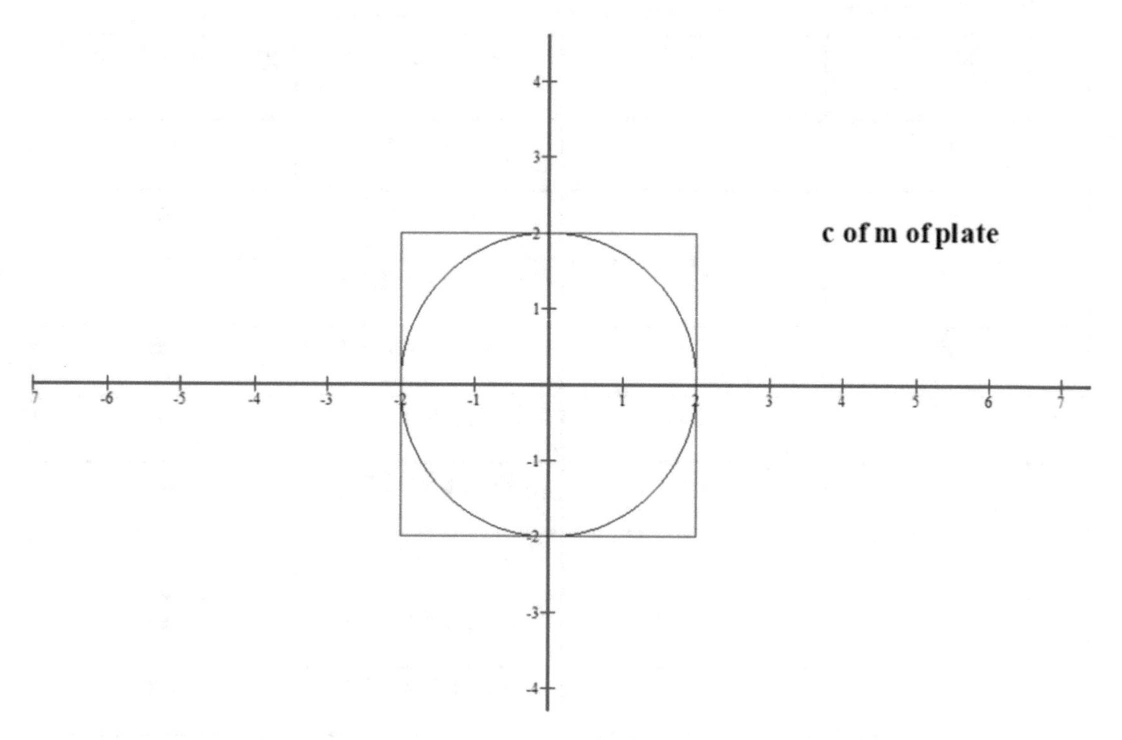

Figure 10-4. *Method of Center of Mass calculation*

We use the random number generator to give us x coordinates between -2 and +2 and y coordinates between -2 and +2. All of our points will be inside the red box as shown earlier. We then use the standard formula for a circle centered on the origin of radius 2 units.

$$\mathbf{x^2 + y^2 = 2^2}$$

We can use this formula to demonstrate manually how the Monte Carlo technique works. Looking at the diagram earlier, we see that our circle is contained within a square. As our random numbers are between -2 and +2 for both x and y, then they all lie within the square. When we generate the numbers, we can test whether they lie inside the circle. We do this by substituting the x and y generated numbers into the preceding formula. If their sum is less than 2^2, then that point lies inside the circle. We can demonstrate this in the following table.

x	y	$x^2 + y^2$	Is $x^2 + y^2 < 4$
0	0	0+0=0	yes
1	1	1+1=2	yes
1.9	1.9	3.61+3.61 = 7.22	no
1.8	1.8	3.24+3.24 = 6.48	no
1.7	1.7	2.89+2.89 = 5.78	no
1.6	1.6	2.56+2.56 = 5.12	no
1.5	1.5	2.25+2.25 = 4.5	no
1.4	1.4	1.96+1.96 = 3.92	yes
1	1.7	1+2.89 = 3.89	yes

Each (x, y) pair represents our generated values and also represents a point on the graph in Figure 10-4. In the third column, we calculate $x^2 + y^2$, and in the fourth column, we say whether the calculated value is less than 2^2. If it is, then it lies within the circle.

In our program, we use the formula $x^2 + y^2 = 4$, but we rearrange it to make y the subject of the formula.

So we get

$$y = \pm\sqrt{(4 - x^2)}$$

When we generate our random numbers for x and y between -2 and +2, we accept the values where

$$y > -\sqrt{(4 - x^2)} \quad \text{and} \quad y < +\sqrt{(4 - x^2)}$$

So these points will lie inside our circle. We add these x and y values into the fields `xcofm` and `ycofm` which accumulate positive and negative values. We divide this by the total number of points to give our coordinates of center of mass.

We can also accumulate the x and y values into `xout` and `yout` and write them to our file to be printed using the Graph package. We can also add our printed value of the center of mass to this graph using "Insert Point Series".

The code for this is shown in the following:

```
/*      cofmc.c
     Center of Mass Calculation.
     Calculates c of m for
     circle center = (0,0) radius = 2
*/
#define _CRT_SECURE_NO_WARNINGS
#include <stdlib.h>
#include <stdio.h>
#include <math.h>
#include <time.h>
double randfunc(); /* Function to return random number */
void main()
{

      int I,outcount;
      float area,total,count;
      FILE *fptr;
      time_t t;
      /* Local arrays */
      double x, y,xout[3500],yout[3500],xcofm,ycofm;

      fptr=fopen("cofmc.dat","w");

   /* Intializes random number generator */
   srand((unsigned) time(&t));

      /* clears arrays to zero */
         for( I = 0; I<3500;I++)
      {
            xout[I] = 0.0;
            yout[I] = 0.0;

      }
      /* Set x and y cofm accumulators to zero */
      xcofm=0.0;
      ycofm=0.0;
```

```
total = 0.0;
count = 0.0;
outcount = 0;
  for( I = 1;I<= 3500;I++)
{
/* Call random number function */

/* Get x values between -2 and +2 */
/* Get y values between -2 and +2 */
      x = randfunc()*4.0-2.0;
      y = randfunc()*4.0-2.0;

/* If the generated x and y values show y is greater than */
/*  - √(4-x^2 ) and less than + √(4-x^2), then add 1 to count */
/* and update the x and y cofm values */

     if(y>-sqrt(4-pow(x,2)) && y<sqrt(4-pow(x,2)))
    {
      xcofm=xcofm+x;
      ycofm=ycofm+y;

      total = total+1;
      outcount = outcount +1;
      xout[outcount] = x;
      yout[outcount] = y;
    }
    count = count+1;

   }
area=(total/count)*16; /* Area is part of the square which is 4x4 or
16 sq units */
printf("total is %f count is %f\n",total,count);

xcofm=xcofm/total;
ycofm=ycofm/total;

printf("area is %lf\n",area);
printf("cofm is %lf,%lf",xcofm,ycofm);
```

```
        /*  Plot the data */

        if(outcount >= 2700)
              outcount = 2700;

           for(I = 1; I<=outcount-1;I++)
              fprintf(fptr,"%lf %lf\n",xout[I],yout[I]);
         fclose(fptr);

}

double randfunc()
{
        /* Get a random number 0 to 1 */
        double ans;

        ans=rand()%1000;
        ans=ans/1000;

         return ans;

}
```

The graph completed is shown in Figure 10-5. The yellow dot shows the center of mass at the center of the plate as we expected. The red dots are all of our points from our random number test.

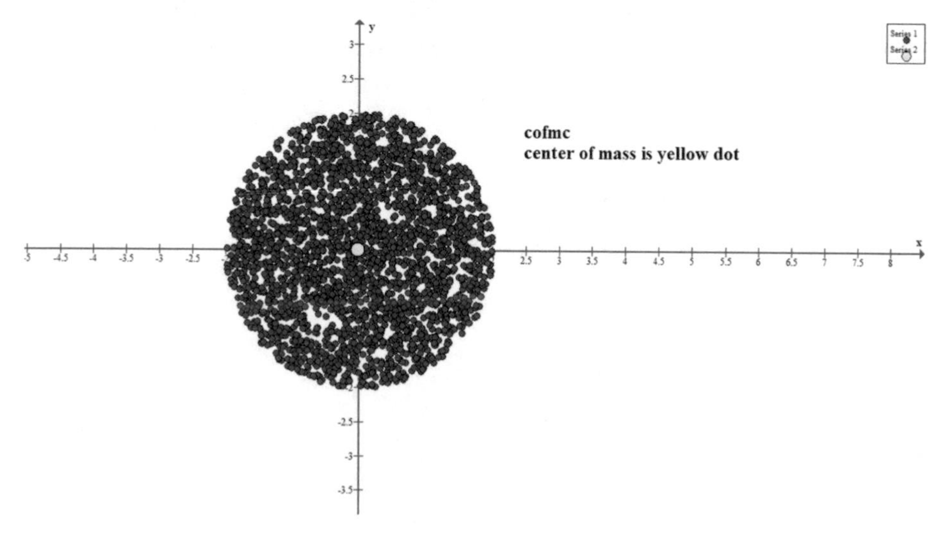

Figure 10-5. *Center of Mass of the plate*

10.3 Other Shapes

In our next example, we want to find the center of mass of a 2D shape. The shape is the curve with the formula $y=x^2$ up to the point where y = 4. This is shown in Figure 10-6.

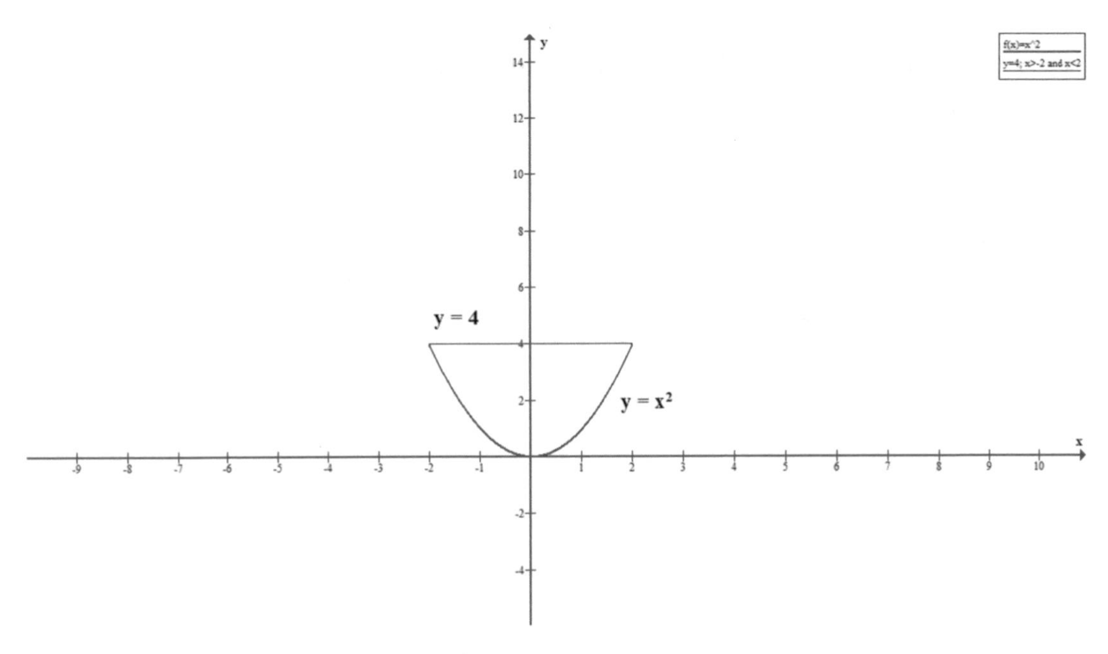

Figure 10-6. *Method for Center of Mass*

The code for this is similar to that for the circle. The main difference is the formula we need to test to decide if we want to accept the randomly generated point. In this case the test is y > pow(x,2). We generate our random points for x between -2 and +2 and for y between 0 and 4.

```
/*      cofm5a.c
      Center of Mass Calculation.
       Calculates c of m for 2D shape y = x2
       below the line y=4
*/
#define _CRT_SECURE_NO_WARNINGS
#include <stdlib.h>
#include <stdio.h>
#include <math.h>
```

```
#include <time.h>
double randfunc();  /* Function to return random number */

void main()
{

       int  I,outcount;
       float area,total,count;
       FILE *fptr;
       time_t t;
       /*  Local arrays */
       double x, y,xout[3500],yout[3500],xcofm,ycofm;

       fptr=fopen("cofm5a.dat","w");

   /* Intializes random number generator */
   srand((unsigned) time(&t));

       /* clears arrays to zero */
          for( I = 1; I<=3500;I++)
       {
              xout[I] = 0.0;
              yout[I] = 0.0;

       }
       /* Set x and y cofm accumulators to zero */
       xcofm=0.0;
       ycofm=0.0;

       total = 0.0;
       count = 0.0;
       outcount = 0;
         for( I = 0;I< 3500;I++)
       {

       /* Call random number function */
```

```
/* Get x values between -2 and +2 */
/* Get y values between 0 and +4 */
            x = randfunc()*4.0-2.0;
            y = randfunc()*4.0;

/* If the generated x and y values are above */
/* the curve y=x2, then add 1 to count */
/* and update the x and y cofm values */

           if(y>pow(x,2))
          {
            xcofm=xcofm+x;
            ycofm=ycofm+y;

            total = total+1;
            outcount = outcount +1;
            xout[outcount] = x;
            yout[outcount] = y;
          }
          count = count+1;

        }

area=(total/count)*16; /* Area is part of the square which is 4x4 or
                          16 sq units */
printf("total is %f count is %f\n",total,count);

xcofm=xcofm/total;
ycofm=ycofm/total;

printf("area is %lf\n",area);
printf("cofm is %lf,%lf",xcofm,ycofm);

/*  Plot the data */

if(outcount >= 2700)
      outcount = 2700;
```

```
        for(I = 1; I<=outcount-1;I++)
            fprintf(fptr,"%lf %lf\n",xout[I],yout[I]);
        fclose(fptr);

}

double randfunc()
{
      /* Get a random number 0 to 1 */
      double ans;

      ans=rand()%1000;
      ans=ans/1000;

       return ans;

}
```

The results are shown in Figure 10-7. The yellow dot shows the center of mass. Both of our programs also output the area of the object being investigated. This is done by counting the number of red dots and taking this as a fraction of the area of the rectangle surrounding it.

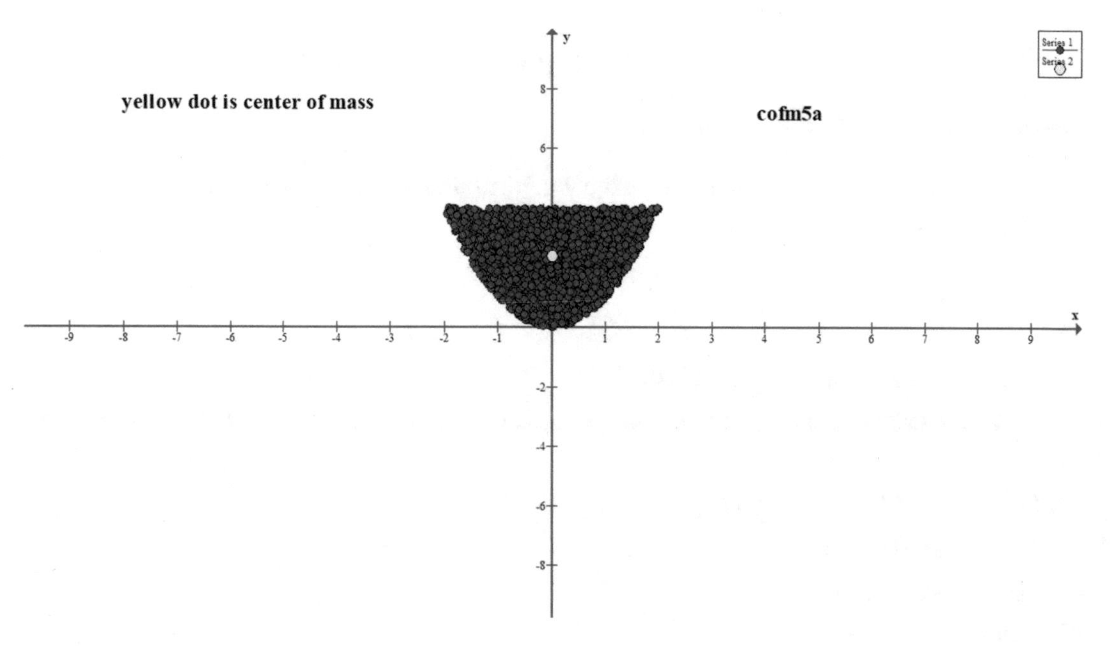

***Figure 10-7.** Center of Mass of shape*

Our final example gives an interesting answer. We want to find the center of mass between the curves $y=x^2$ and $y=x^2 + 1$ and the line y=4. The following diagram, in Figure 10-8, shows this in a graph.

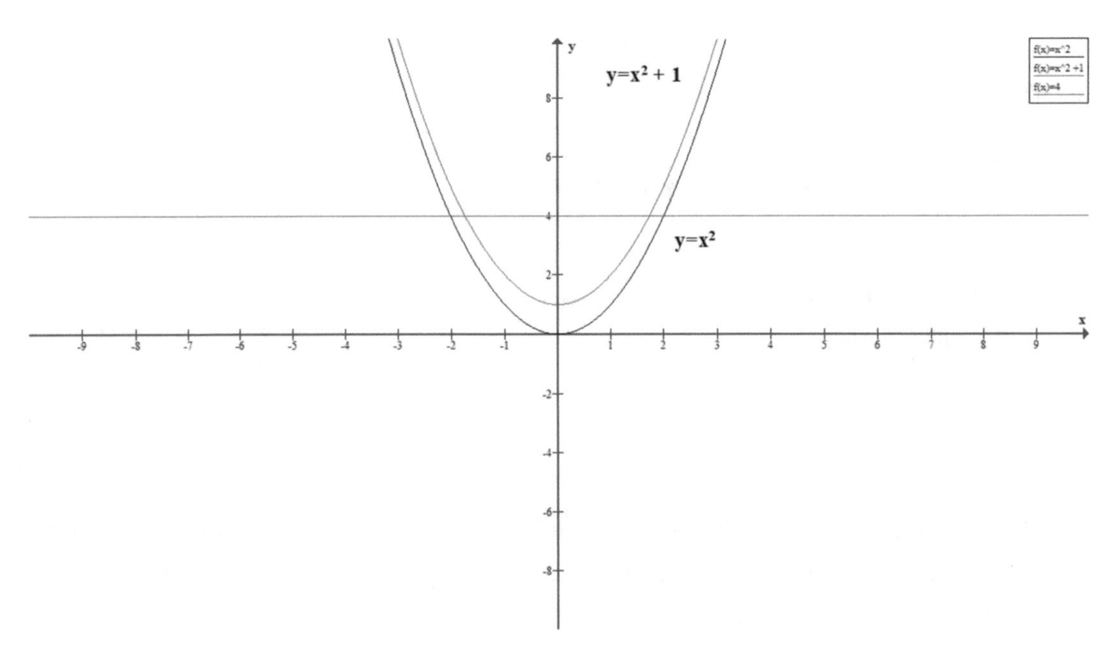

***Figure 10-8.** Center of Mass method*

Our main code for the inspection of accepting the generated points is

```
y>pow(x,2) && y<(pow(x,2)+1)
```

so our y values will be above the curve $y=x^2$ and below the curve $y=x^2 + 1$.

The code is shown as follows:

```
/* cofm5b.c */
/*
     Center of Mass Calculation.
      Calculates c of m for 2D shape between y = x2, y = x2 + 1, and y = 4
*/
#define _CRT_SECURE_NO_WARNINGS
#include <stdlib.h>
#include <stdio.h>
#include <math.h>
```

```
#include <time.h>
double randfunc();/* Function to return random number */

void main()
{

      int  I, outcount;
      float area,total,count;
      FILE *fptr;
      time_t t;
      /*  Local arrays */
      double x, y,xout[3500],yout[3500],xcofm,ycofm;

      fptr=fopen("cofm5b.dat","w");

   /* Intializes random number generator */
   srand((unsigned) time(&t));

       /* clears arrays to zero */
         for( I = 1; I<=3500;I++)
      {
            xout[I] = 0.0;
            yout[I] = 0.0;

      }
      /* Set x and y cofm accumulators to zero */
      xcofm=0.0;
      ycofm=0.0;

      total = 0.0;
      count = 0.0;
      outcount = 0;
        for( I = 0;I< 3500;I++)
      {
      /* Call random number function */

      /* Get x values between -2 and +2 */
      /* Get y values between 0 and +4 */
            x = randfunc()*4.0-2.0;
            y = randfunc()*4.0;
```

```
    /* If the generated x and y values are above */
    /* the curve y=x2 and below y=x2+ 1, then add 1 to count */
    /* and update the x and y cofm values */

         if(y>pow(x,2) && y<(pow(x,2)+1))
        {
          xcofm=xcofm+x;
          ycofm=ycofm+y;

          total = total+1;
          outcount = outcount +1;
          xout[outcount] = x;
          yout[outcount] = y;
        }
        count = count+1;

       }
    area=(total/count)*16; /* Area is part of the square which is 4x4 or
                              16 sq units */
    printf("total is %f count is %f\n",total,count);

    xcofm=xcofm/total;
    ycofm=ycofm/total;

    printf("area is %lf\n",area);
    printf("cofm is %lf,%lf",xcofm,ycofm);

    /* Plot the data */

    if(outcount >= 2700)

          outcount = 2700;

       for(I = 1; I<=outcount-1;I++)
          fprintf(fptr,"%lf %lf\n",xout[I],yout[I]);
     fclose(fptr);

}
```

```
double randfunc()
{
        /* Get a random number 0 to 1 */
        double ans;

        ans=rand()%1000;
        ans=ans/1000;

            return ans;
 }
```

The following, Figure 10-9, is the graph showing all of the accepted random points and the position of the center of mass. The blue dot is the position of the center of mass. For this boomerang-shaped curve, the center of mass is outside the object!

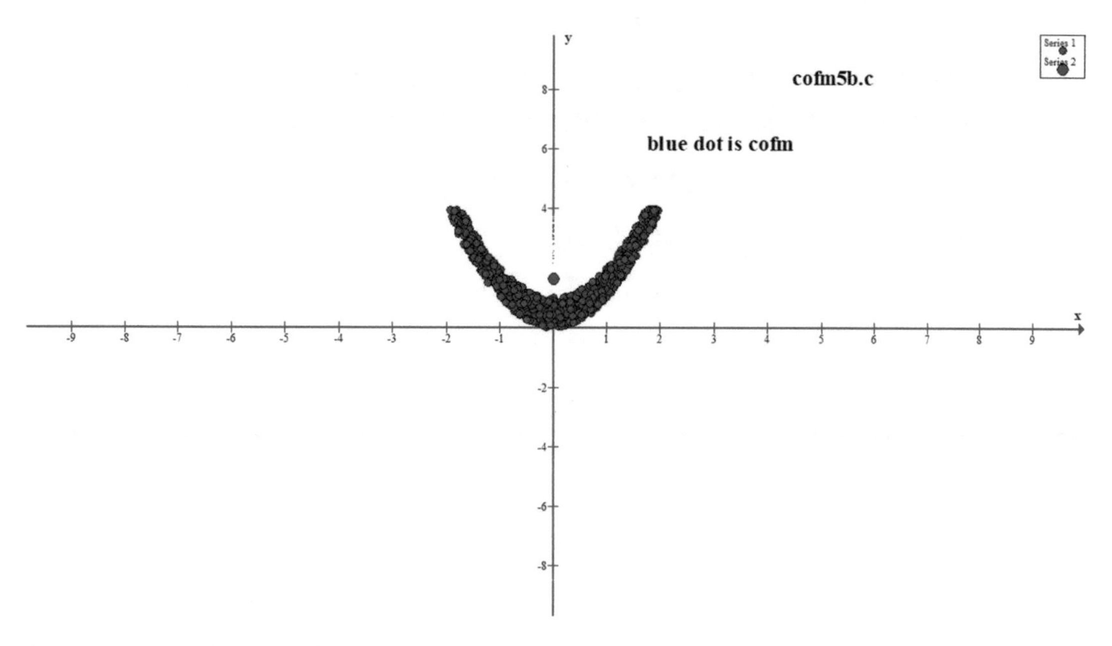

***Figure 10-9.** Center of Mass of boomerang*

EXERCISES

1. Amend any of the earlier programs to find the center of mass of the oval-shaped plate shown at the start of this chapter.

   The general equation of an ellipse centered on the origin is

   $$(x^2 / a^2) + (y^2 / b^2) = 1$$

   If you take the oval plate to be an ellipse and to be centered on the origin with a=2 and b=1, then its equation is

   $$(x^2/2^2) + y^2/1^2 = 1$$

   We can solve this for y to get

   $$y = \pm\sqrt{(1 - x^2/4)}$$

2. Amend any of the preceding programs to find the center of mass of the two concentric circles which are centered on the origin. Their formulas are

   $$x^2 + y^2 = 4 \quad \text{and} \quad x^2 + y^2 = 1$$

   The solutions of these equations for y are

   $$y = \pm (4 - x^2)^{1/2}$$

   and

   $$y = \pm (1 - x^2)^{1/2}$$

CHAPTER 11

Brownian Motion

11.1 Brownian Motion Theory

If you could look at one molecule of a gas in the middle of a container of the gas, then the one molecule will move in random directions if the temperature, pressure, and volume are constant throughout the gas.

One way to picture this from a real-life situation is if you imagine the smoke coming from a cigarette. Once the smoke gets about 6 inches above the cigarette (before this the smoke will be warm and therefore rising), then the smoke seems to dart about in random directions.

In 1827 scientist Robert Brown used a microscope to observe pollen moving in water. Albert Einstein later showed that the pollen was being moved by individual water molecules. This was one of the first pieces of scientific evidence for the existence of molecules.

In our program we will assume that an individual molecule is in constant collision with molecules of a gas. As a result, our test molecule can move in any direction. We will restrict this to 2D so that we can plot the resulting graphs. We can use our Monte Carlo methods again for this modeling.

We can use our random number generator to produce a random angle. As the molecule can move in any direction, then the angle can be 0 to 360 degrees or 0 to 2π radians. In order to plot our graph of the movements of the molecule, we will use the angle to produce cosine and sine of the angle. If we simplify the motion of the molecule by saying that after each collision the molecule moves 1 unit of length, we can simplify our calculations.

P. Joyce, *Practical Numerical C Programming*, https://doi.org/10.1007/978-1-4842-6128-6_11

The following diagram, Figure 11-1, shows what we will do.

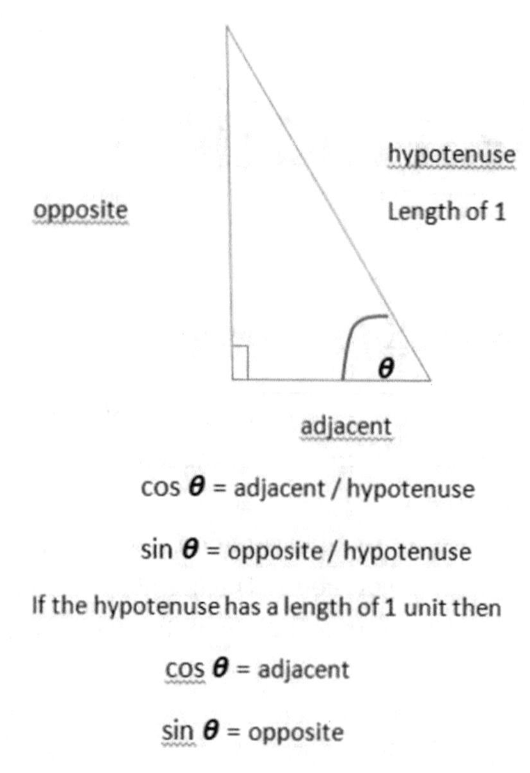

Figure 11-1. *Mechanism for finding x and y moves*

We see the definitions of cos and sin. If, as we said, the distance after each collision is 1 unit of length, then the hypotenuse will be 1, so the cos of the angle will be the "adjacent" of our triangle and the sin of the angle will be the "opposite". So, from the diagram, the adjacent is in the x direction and the opposite is in the y direction.

As our angle generated can be between 0 and 360 degrees, then our cos and sin values can be negative. See the following diagrams, Figure 11-2, for cos and sin.

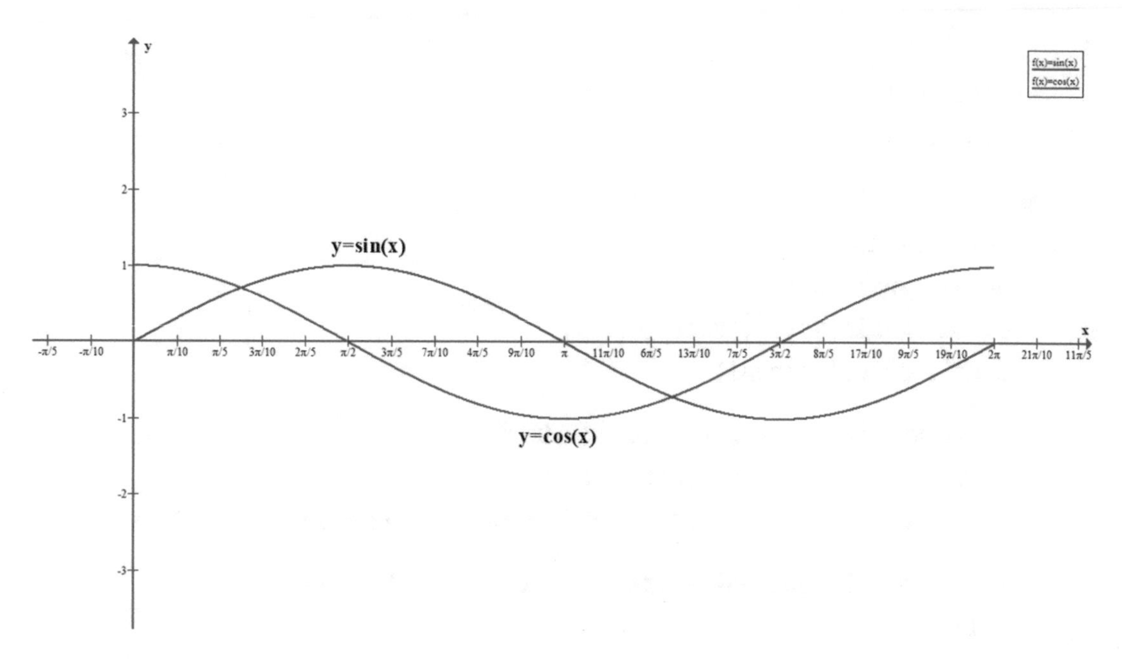

Figure 11-2. *Sin and cos properties from 0 to 2π*

So a negative cos would be a movement in the negative x direction and a negative sin would be a movement in the negative y direction.

In our program we write the (x, y) position of the molecule to the file after each iteration of our Monte Carlo forloop.

As the random number generated by the program will be between 0 and 1, we multiply this by 2π to give us a random number between 0 and 2π.

The code is shown as follows:

```
/* Brownian motion (2D) simulation (Monte Carlo)*/
/* */
#define _CRT_SECURE_NO_WARNINGS
#include <stdlib.h>
#include <stdio.h>
#include <math.h>
#include <time.h>

#define PI 3.141592654

void main()
```

```
{
      FILE *fptr;
      time_t  t;

int i;
      int collisions;
      double anglerand;
      double xvals[5950],yvals[5950];
      double cosval,sinval;

      /* Create and open our output file */
      fptr=fopen("browntest3.dat","w");
      srand((unsigned) time(&t)); /* Set the random number seed */

      /* Preset the variables used in the calculation */
      collisions=1000;
      xvals[0]= 0.0;
      yvals[0]= 0.0;

      for(i=0;i<1000;i++)
      {

            anglerand=rand()%1000;
            anglerand=anglerand/1000;

            /* Get a random angle between 0 and PI radians */

            anglerand=2*PI*anglerand;

            /* Length of jump is 1 */
            /* So the cos and sin of the angle */
            /* will be the distance moved in */
            /* that direction (+ or -) */

            xvals[i+1]=xvals[i]+cos(anglerand);
            cosval = cos(anglerand);
```

```
            yvals[i+1]=yvals[i]+sin(anglerand);
            sinval = sin(anglerand);

            /* Print the current x and y values to the file */
            fprintf(fptr,"%lf %lf\n", xvals[i], yvals[i]);

            /* This can be used in the program to print */
            /* the current cos and sin values if required */
            /*printf("cosval = %lf sinval = %lf\n",cosval,sinval);*/

        }

}
```

The results are shown in Figure 11-3. The three different colors represent three different runs of the program. You can see the three different general directions that our test molecule took for each run. If you run the program three times yourself, your three graphs will be different to these but will show the same general form.

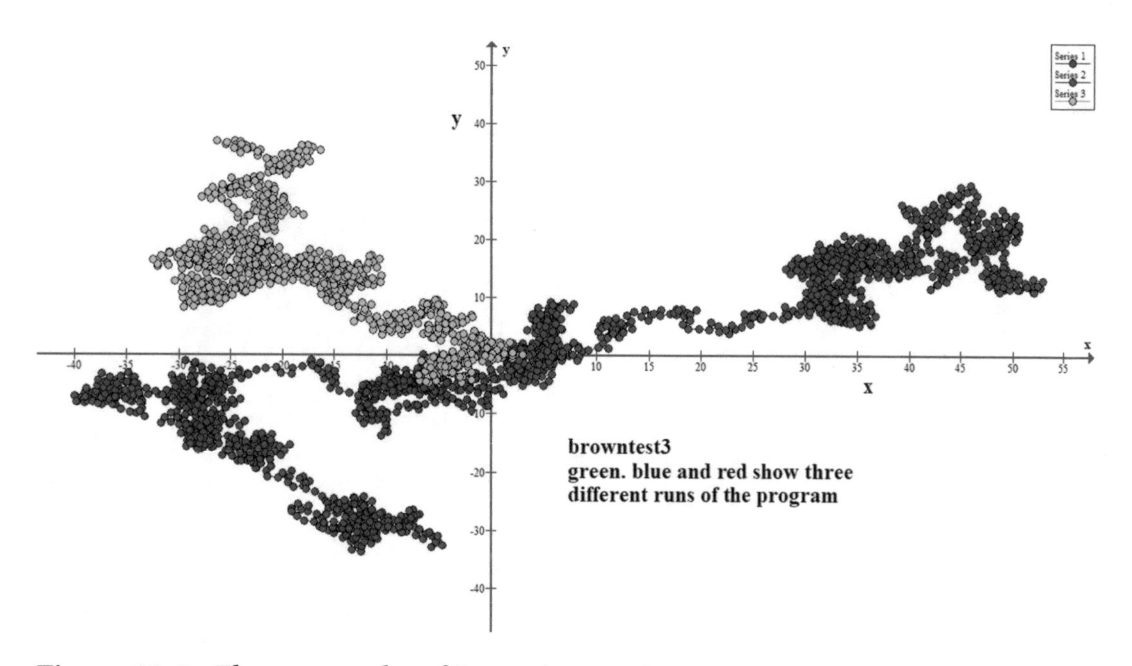

Figure 11-3. *Three examples of Brownian motion*

EXERCISES

1. Write a program to do the same as the program in this chapter except that instead of choosing a random angle, you choose random values between 0 and 1 for x and y movements. You will also need to choose random + or – moves.

CHAPTER 12

Diffusion Lattice Model

12.1 Vacancy Lattice Diffusion

This is a simplified model of the structure of atoms in solids which allows us to demonstrate their movement. There are different types of diffusion. The type we will be looking at is vacancy diffusion.

Different solids have different molecular structure, and some of the more common structures have a name, for example, "primitive cubic," "face-centered cubic," and "body-centered cubic."

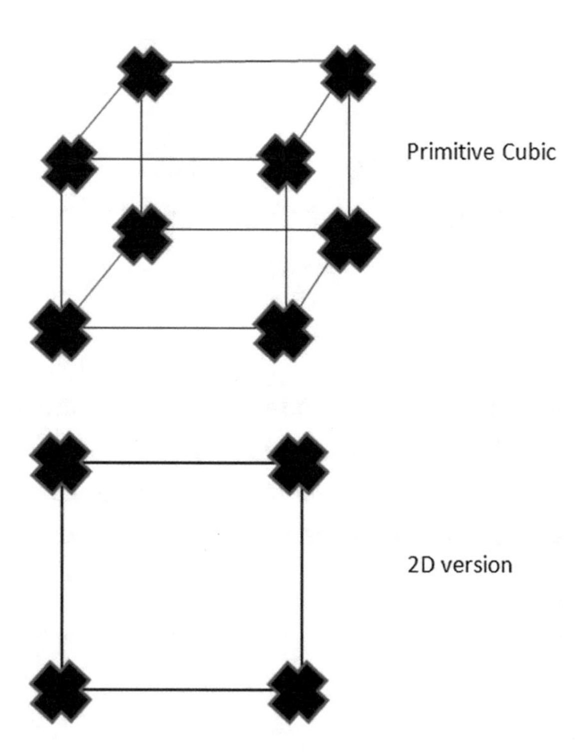

Figure 12-1. *Primitive cube and 2D equivalent*

P. Joyce, *Practical Numerical C Programming*, https://doi.org/10.1007/978-1-4842-6128-6_12

The Primitive cubic is shown as the upper diagram in Figure 12-1. This is a cube with an atom at each corner of the cube. The face-centered cubic structure is like the Primitive cubic but with an additional atom in the center of each face of the cube. The body-centered cubic structure is like the Primitive cubic but with one extra atom in the center of the cube.

We will be concentrating on the Primitive cubic structure. The 2D case of the Primitive cube is just a square with an atom at each corner. For a solid with a Primitive cubic structure, the atoms at the corners are shared with neighboring Primitive structures. This can also translate to the 2D case as shown in Figure 12-2.

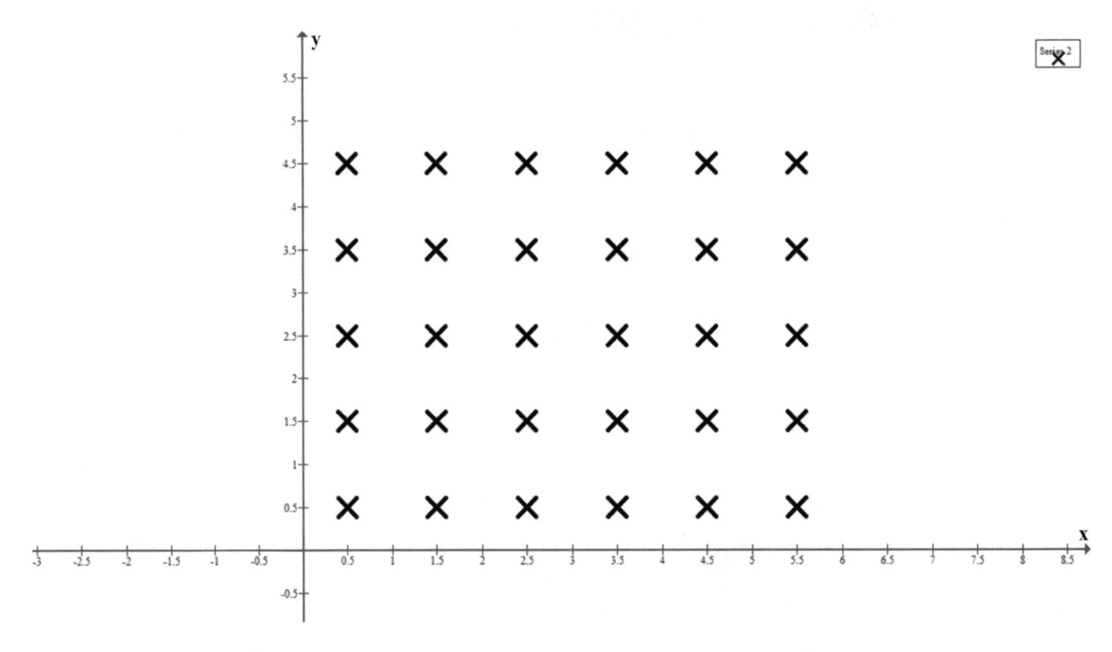

Figure 12-2. *2D lattice*

The four atoms making up the Primitive "square" structure in the bottom-left corner of the diagram correspond to the 2D Primitive diagram previously. The top two atoms in this case are the bottom two atoms of the preceding structure. Similarly, the two atoms on the right of the structure are shared as the left-hand two atoms of this structure.

There are different types of diffusion in atoms. In our example, we will assume that other atoms can move within the structure shown earlier. These atoms can move in the spaces between the atoms of the structure. This means that they can only move up, down, right, or left.

The following diagram, Figure 12-3, shows the lattice as the X characters and the possible positions as the red dots. The atom which will move is the gray dot.

In order to manage what happens when the atom gets to the edge of the lattice, we will assume that it "bounces" back into the lattice.

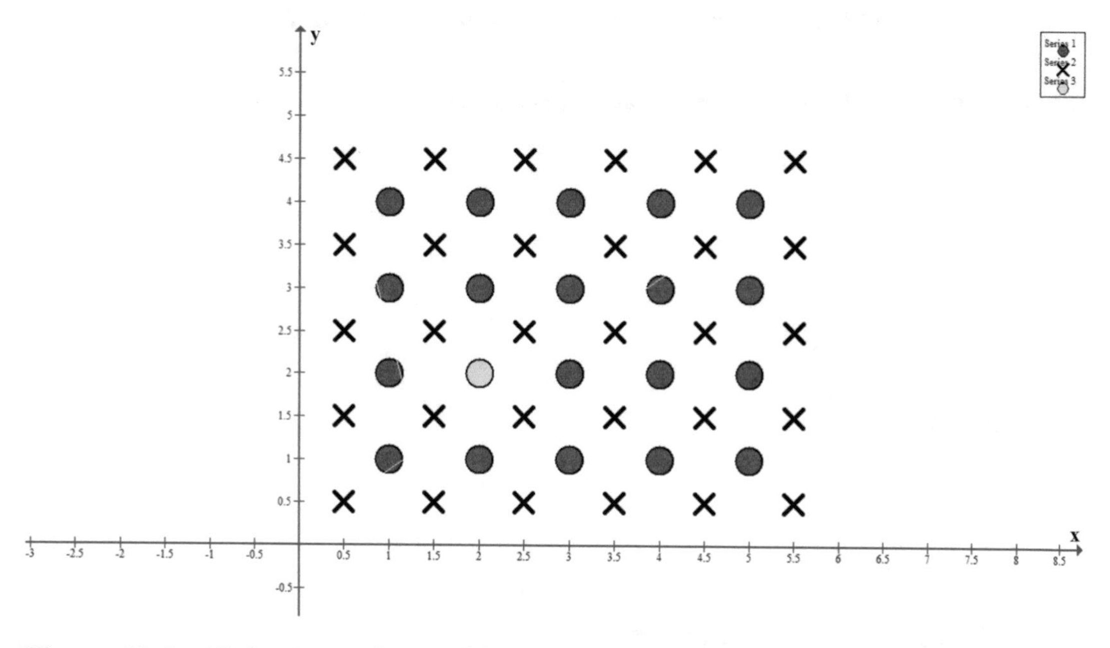

Figure 12-3. *2D lattice with possible moves*

In the following program, we assume that the lattice is a 20X20 structure.

We have two random number generator functions. One, `int IRND();`, is used to select a random integer between 0 and 19 if we want the start position of the atom to be in a random place. The other random generator function, `int IFOURRND();`, returns integers 1, 2, 3, or 4. Here, 1 indicates that we want the atom to move to the right, 2 indicates a move up, 3 indicates a move to the left, and 4 is a move down, so that the direction the atom moves is random each time.

In this program we have commented out the calls to the random function which sets the initial position of the atom as we want to set this position ourselves. This can be changed later. We have set the number of Monte Carlo Cycles, MCCMAX, to 50.

```
/*      PROGRAM vaca.c
        VACANCY DIFFUSION MODEL. (2D VERSION)
*/
#define _CRT_SECURE_NO_WARNINGS
#include <stdlib.h>
#include <stdio.h>
#include <math.h>
#include <time.h>

int IRND();/*Function to return random number from 0 to 19 */

int IFOURRND();/*Function to return random number 1, 2, 3, or 4 */

void main()
{

      int N1,N2;
      int N1N,N2N,MCC;
      int LATTICE2[20][20];
      int MCCMAX;
      int Q,P,INC;
      FILE *fptr;

      time_t t;

      /* Intializes random number generator */
      srand((unsigned) time(&t));
      fptr=fopen("vacxa.dat","w");

      MCCMAX = 50; /* Set number of Monte Carlo Cycles */

      for(P=0;P<20;P++)
      {
            for(Q=0;Q<20;Q++)
            {
            /* FILL THE ARRAY */

                  LATTICE2[P][Q] = 0;
```

```
        }
}

/* SELECT ANY SITE AS THE INITIAL VACANCY SITE*/
/* Can be set randomly using the IRND function */
/* or can be set to specific values */

/*N1N = IRND(); Start x and y values if you want
N2N = IRND(); random positions (commented out in this case)*/
N1N=1; /* Start x value */
N2N=10; /* Start y value */

LATTICE2[N1N][N2N] = 1; /* Set vacancy site in lattice */

/* Monte Carlo Cycle loop */
/* Loops round MCCMAX number of times */

for(MCC=1;MCC<=MCCMAX;MCC++)
{

        N1=N1N; /* Set N1 (current lattice x value) */
        N2=N2N; /* Set N2 (current lattice y value) */

        if(LATTICE2[N1][N2] == 1)
        {
                /* VACANCY SITE (= 1 )*/

                /* Call function to randomly select 1, 2, 3, or 4 */

                INC = IFOURRND();
                /* 1 indicates a move to the right */
                /* 2 indicates a move up */
                /* 3 indicates a move to the left */
                /* 4 indicates a move down */

                /* Instead of going from 19 to 1, etc., you bounce off
                the boundary. Go from 19 to 18, etc. */
```

```
                if(INC == 1 ) /* right */
                {
                      if(N1 == 19)
                            N1N = 18;
                      else
                            N1N = N1+1;

                      }else if(INC == 2)  /* Up */
                      {
                            if(N2 == 19)
                                  N2N = 18;
                            else
                                  N2N = N2+1;

                      }else if(INC == 3) /* Left */
                      {
                            if(N1 == 1)
                                  N1N = 2;
                            else
                                  N1N = N1-1;

                      }else if(INC == 4) /* Down */
                      {
                            if(N2 == 1)
                                  N2N = 2;
                            else
                                  N2N = N2-1;

                      }

                      if(LATTICE2[N1N][N2N] == 0)
                      {
                           LATTICE2[N1N][N2N] = 1; /* Set as a used site */

                      }
                            else
                                  printf("not found\n");
                      }

      }
```

```
    /* Write any used lattice positions to file */
    for(P=0;P<20;P++)
    {
        for(Q=0;Q<20;Q++)
        {

            if(LATTICE2[P][Q] == 1)
                fprintf(fptr," %d\t%d\n",P,Q);

        }
    }

    fclose(fptr);

}

int IRND()
{
    /* Generate a random whole number from 0 to 19 */
    double TOT,DIV,X;
    int ANS,I;

    TOT=rand()%1000;
    TOT=TOT/1000;

    /* Returns 0,1,2 ... or 19 */
    /* chosen at random */

    DIV = 20.0;
    X = 1.0;
    for(I=0;I<20;I++)
        if(TOT < X/DIV)
            ANS = I;

        else
            X = X+1.0;

    return ANS;
}
```

```
int IFOURRND()
{
        /* Generate a random whole number 1, 2, 3, or 4 */
        double TOT;
        int ANS;

        TOT=rand()%1000;
        TOT=TOT/1000;

        /* Returns 1, 2, 3 or 4 */
        /* chosen at random */

        if(TOT < 0.25)
                ANS = 1;
        else if(TOT < 0.5)
                ANS = 2;
        else if(TOT < 0.75)
                ANS = 3;
        else
                ANS = 4;

        return ANS;

}
```

After each iteration of the loop, we store the current position of the atom and we save these into the file vacxa.dat.

The output from the file is shown in Figure 12-4. This is for 50 Monte Carlo Cycles starting at lattice position (1,10).

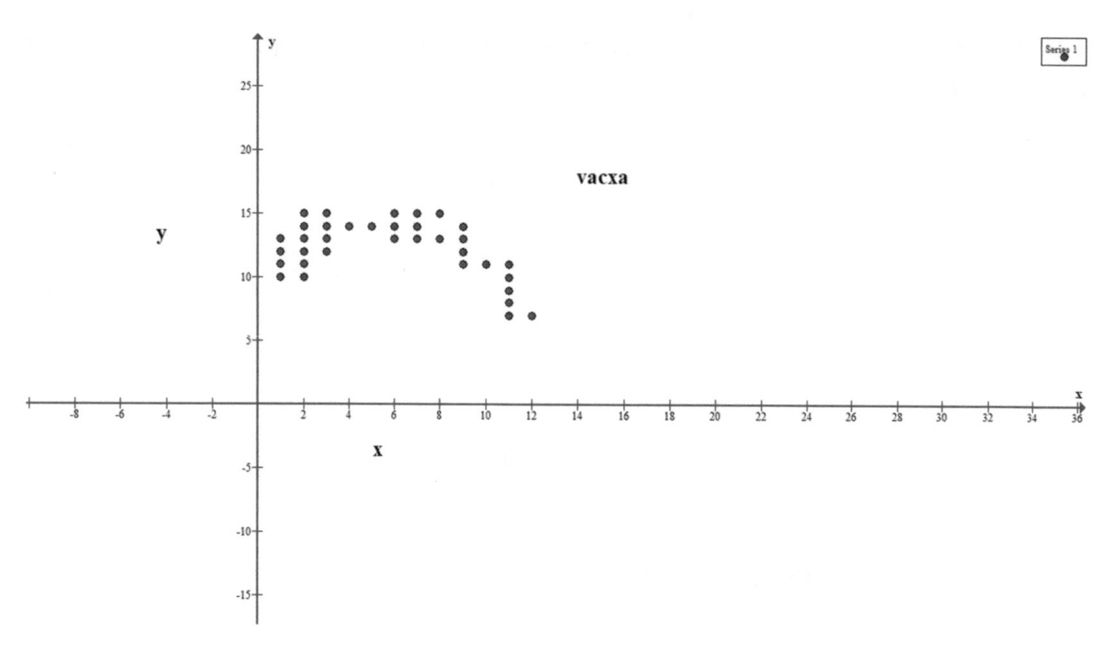

Figure 12-4. *2D lattice MCC=50 start point (1,10)*

The next graph, Figure 12-5, shows the lattice after 250 Monte Carlo Cycles starting at lattice position (1,10).

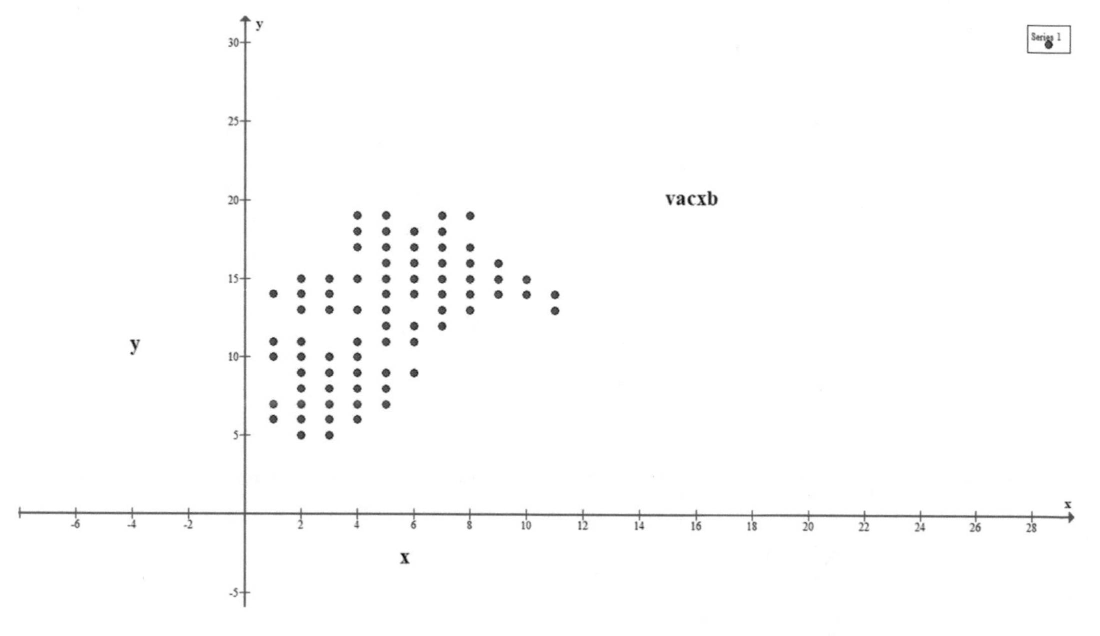

Figure 12-5. *2D lattice MCC=250 start point (1,10)*

The next graph, Figure 12-6, shows the lattice after 100 Monte Carlo Cycles starting at lattice position (19,1).

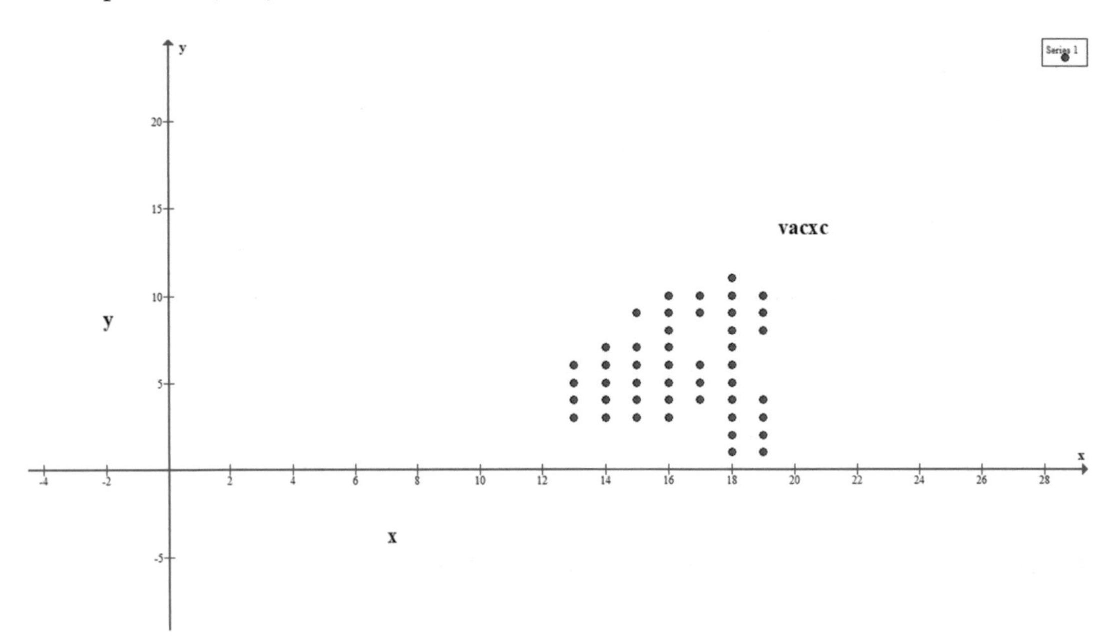

Figure 12-6. *2D lattice MCC=100 start point (19,1)*

The next graph, Figure 12-7, shows the lattice after 50 Monte Carlo Cycles starting at lattice position (10,10).

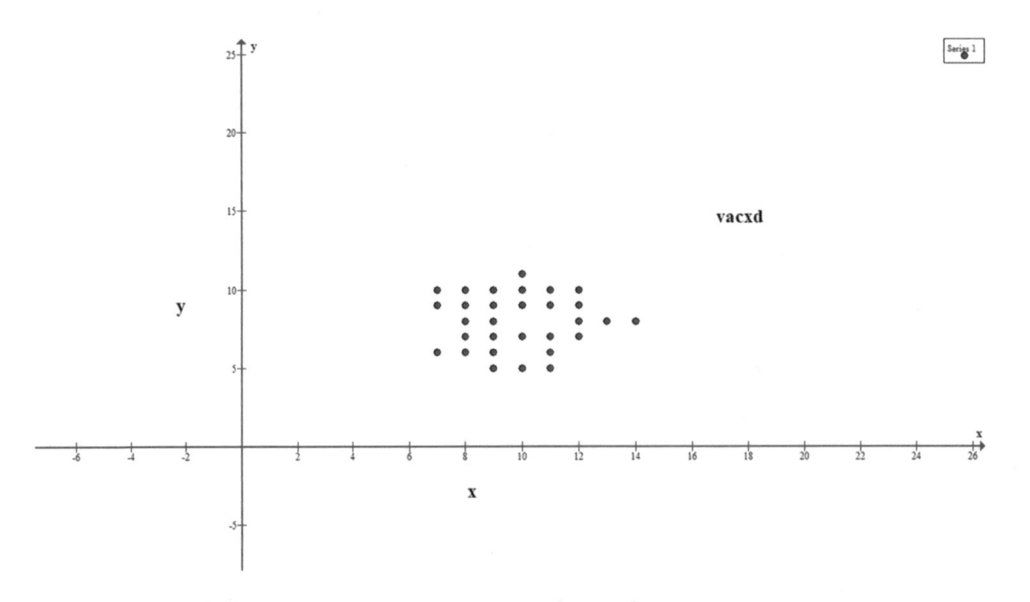

Figure 12-7. *2D lattice MCC=50 start point (10,10)*

The next graph, Figure 12-8, shows the lattice after 500 Monte Carlo Cycles starting at lattice position (10,10).

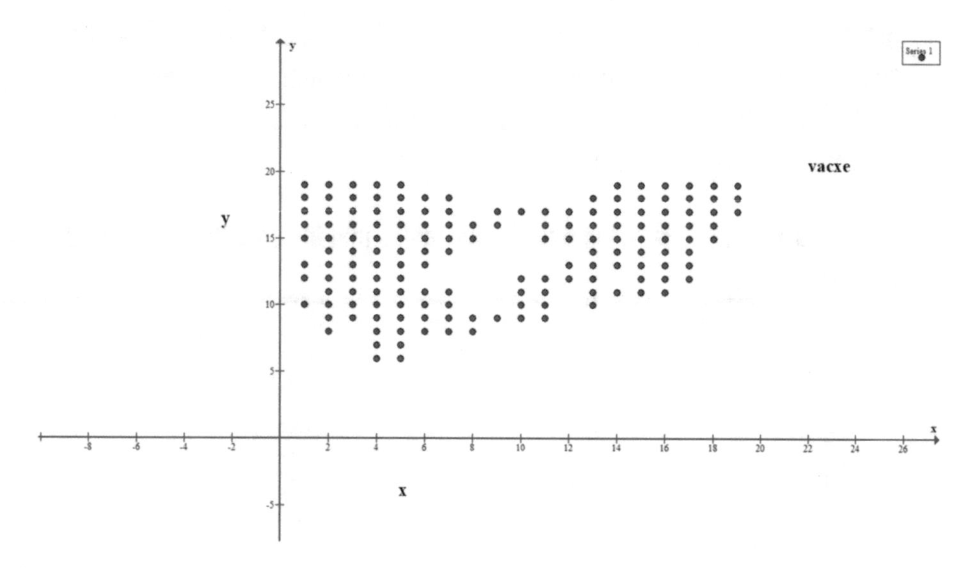

Figure 12-8. *2D lattice MCC=500 start point (10,10)*

The next graph, Figure 12-9, shows the lattice after 10000 Monte Carlo Cycles starting at lattice position (10,10).

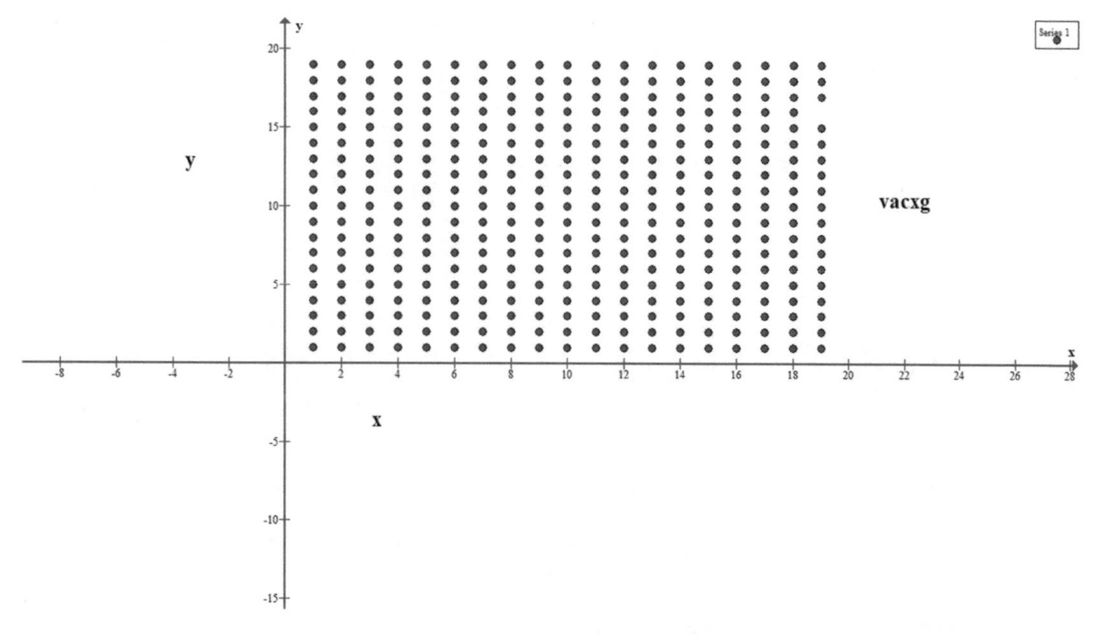

Figure 12-9. *2D lattice MCC=10000 start point (10,10)*

EXERCISES

1. Amend the vaca.c program so that the lattice is divided into two sections at the line x=9. The division is an obstruction blocking movement of the atoms from right to left or left to right. In the middle of the obstruction is a hole, say at x=9, y=9. Atoms can pass through this hole.

   Run the program and print the graph of the lattice that the program produces. Set MCCMAX to 1000.

CHAPTER 13

Chain Reaction

13.1 Chain Reaction Theory

If you are familiar with the periodic table of elements, you will know that the elements in the table are listed in the order of the number of protons in their nucleus. So the first element is hydrogen whose nucleus contains 1 proton and the last element of the naturally occurring elements is uranium which contains 92 protons.

As well as containing protons, the nucleus also contains neutrons. However, in the case of neutrons, the nucleus can contain a varying number. This is what gives rise to the existence of "isotopes." The heavier elements contain more neutrons than protons. In the case of uranium, one of the isotopes is uranium-235. The notation of this is ^{235}U. The number 235 just refers to the sum of the number of protons and the number of neutrons in the nucleus. As the number of protons for uranium has to be 92, then the number of neutrons in uranium-235 must be 235 - 92 = 143. The notation for this is ${}_{92}{}^{235}U$.

These "heavy" nuclei tend to break up into nuclei with fewer protons and neutrons, and so these "broken" nuclei must be other elements from the periodic table.

$${}_{92}{}^{235}U \rightarrow {}_{90}{}^{231}Th + {}_{2}{}^{4}He$$

The preceding diagram shows the decay of uranium-235 into thorium-231 and helium. The way of representing the nuclei in this type of diagram is to put the atomic weight to the top left of the symbol of the element and its atomic number (number of protons) to the bottom left. We can see from the diagram that the atomic weights of thorium and helium, when added, make 235, the atomic weight of the original uranium nucleus. A similar process is used for the atomic numbers.

P. Joyce, *Practical Numerical C Programming*, https://doi.org/10.1007/978-1-4842-6128-6_13

When these elements break up, as well as forming new elements, they can also eject neutrons which fly off at high speed. If one of these neutrons hits a nucleus of uranium-235, it could produce the reaction shown as follows:

$$_{92}^{235}U + _{0}^{1}n \rightarrow _{92}^{236}U \rightarrow _{54}^{140}Xe + _{38}^{93}Sr + 3_{0}^{1}n$$

where $_{0}^{1}n$ just represents a neutron

$_{92}^{236}U$ is uranium-236 as it has absorbed the incoming neutron into its nucleus

$_{54}^{140}Xe$ is xenon-140

$_{38}^{93}Sr$ is strontium-93

$3_{0}^{1}n$ is 3 neutrons

This is pictured in Figure 13-1.

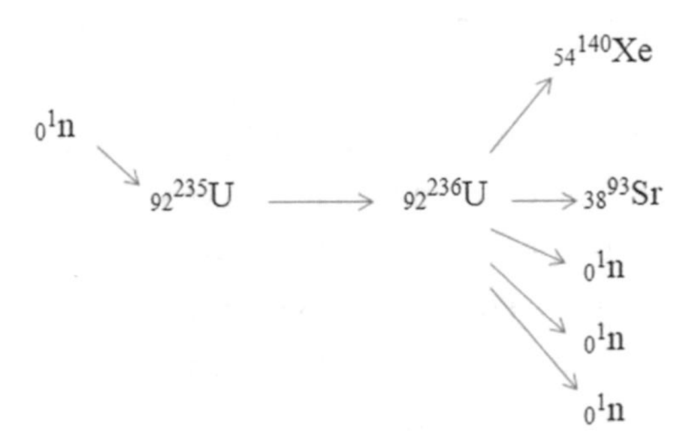

Figure 13-1. *Schematic diagram of decay of uranium-236*

So in this case, the uranium-236 decays into two different nuclei but also emits three neutrons. If these three neutrons go on to collide with three more uranium-236 nuclei, then each of these will give off three neutrons making nine neutrons in all. If these then go on to collide with nine more uranium-236 nuclei, we will get 3 x 9 = 27 neutrons emitted. So you can see how this would create a "cascade" of neutrons. This is the basis of a "chain reaction." The breakup of nuclei is called "fission."

When neutrons are emitted from these collisions, they move away at high speed. If the solid containing the uranium is small, then the neutrons will escape from it before they can meet another uranium nucleus, so the chain reaction will not take place. So the size of the solid is crucial. This size is called the "critical mass." What is also important is the shape of the solid. The size and shape of our solid that we need for a chain reaction to take place is what we will investigate in this chapter.

The following diagram, Figure 13-2, shows our block of material. It will have a square cross-section of length a and a side of length b.

We will start with a nucleus shown by the blue dot at position (x_0, y_0, z_0).

We will assume that two neutrons are emitted as shown. One moves to position (x_1, y_1, z_1) and the other moves to position (x_2, y_2, z_2).

We can use the Monte Carlo method of generating random numbers for the initial and final positions of the neutrons. If the final positions are outside of the block, then there will be no chain reaction from these neutrons. We now generate another initial position and see where the resulting neutrons move to. We keep a count of the number of neutrons which stay within the block. The number of these as a fraction of the number of original nuclei is called the "survival fraction."

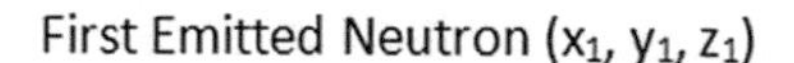

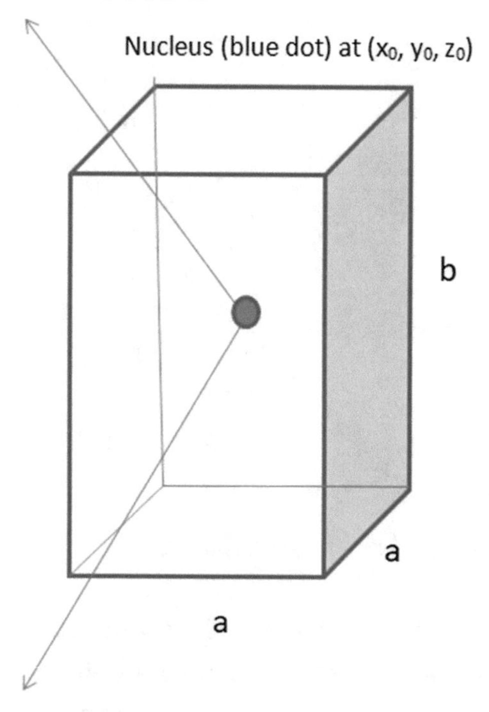

Figure 13-2. *Block to test movement of decay neutrons*

We can use the formula $f = N_{in} / N$, where f is the survival fraction, N_{in} is the number of neutrons that stay within the block, and N is the number of fissions we test.

13.2 Chain Reaction Program

In our first program, we will use blocks with varying values of a and b, and we will test the significance of the shape of the block. We can then use b/a as an indicator of the shape of the block and plot this against the survival fraction f. A survival fraction above 1 is what we are looking for.

The user can enter the number of fissions. This is usually between 100 and 1000.

The program uses spherical coordinates to find the positions of the emitted neutrons.

The following diagram shows this (Figure 13-3).

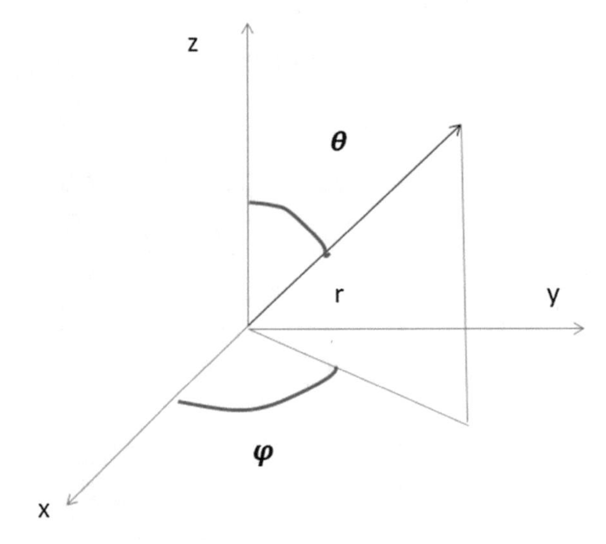

Figure 13-3. *Spherical polar coordinates*

We want to convert our spherical polar coordinates into normal Cartesian coordinates. The following diagram, Figure 13-4, shows how we do this.

The nucleus is at o and an emitted neutron ends up at (x_1, y_1, z_1).

We can project this point onto the x axis, y axis, and z axis as shown. We have just created three right-angled triangles. So the distance moved by the neutron in the x direction is ox_1, the distance moved in the y direction is oy_1, and the distance moved in the z direction is oz_1.

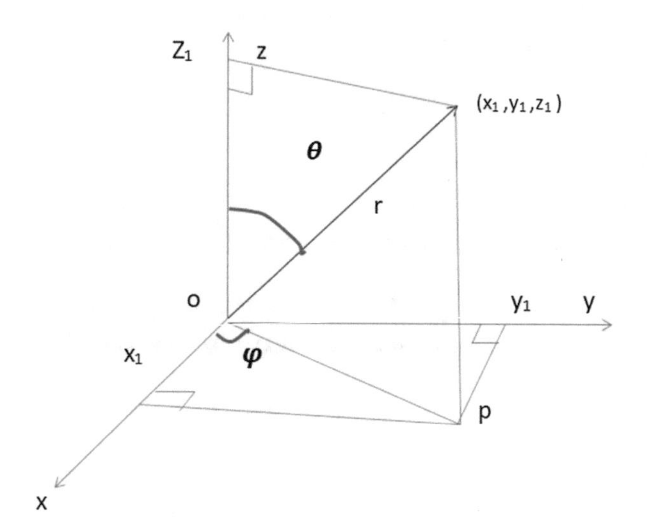

Figure 13-4. *Transformation method to Cartesian*

We can see that $\cos\boldsymbol{\theta} = z_1 / r$, so $z_1 = r \cos\boldsymbol{\theta}$.

Also $\sin\boldsymbol{\theta} = op / r$, so $op = r \sin\boldsymbol{\theta}$.

And $\sin\boldsymbol{\varphi} = y_1 / r \sin\boldsymbol{\theta}$, so $y_1 = r \sin\boldsymbol{\theta} \sin\boldsymbol{\varphi}$.

And $\cos\boldsymbol{\varphi} = x_1 / r \sin\boldsymbol{\theta}$, so $x_1 = r \sin\boldsymbol{\theta} \cos\boldsymbol{\varphi}$.

So $(x_1, y_1, z_1) = (r \sin\boldsymbol{\theta} \cos\boldsymbol{\varphi}, r \sin\boldsymbol{\theta} \sin\boldsymbol{\varphi}, r \cos\boldsymbol{\theta})$.

So our new positions for the two neutrons are given by

$x_1 = x_0 + r \sin\boldsymbol{\theta} \cos\boldsymbol{\varphi}$ $y_1 = y_0 + r \sin\boldsymbol{\theta} \sin\boldsymbol{\varphi}$ $z_1 = z_0 + r \cos\boldsymbol{\theta}$

$x_2 = x_0 + r \sin\boldsymbol{\theta} \cos\boldsymbol{\varphi}$ $y_2 = y_0 + r \sin\boldsymbol{\theta} \sin\boldsymbol{\varphi}$ $z_2 = z_0 + r \cos\boldsymbol{\theta}$

with different values of $\boldsymbol{\theta}$, $\boldsymbol{\varphi}$, and r for the two emitted neutrons. In our program, we use d1 and d2 in the program for the distances traveled by each neutron.

So we generate our Monte Carlo random values of the distance r and the angles $\boldsymbol{\theta}$ and $\boldsymbol{\varphi}$. We generate random values of $\cos\boldsymbol{\theta}$ and then use the arccos function to get the corresponding value of $\boldsymbol{\theta}$.

We adjust the values of b and a after each loop so that we can monitor how the shape of the block affects the survival fraction.

We can plot our graph for the b/a and f relationship after we run the program.

The code for this is as follows:

```
/*      chain1.c
        Chain Reaction Simulation. Cubic Shape
        Predefined values of a & b (varied values    of a & b) */
```

```
#define _CRT_SECURE_NO_WARNINGS
#include <stdlib.h>
#include <stdio.h>
#include <math.h>
#include <time.h>
double randfunc();/* Function to return random number ( 0 to 1 ) */

int checkin(double x, double y, double z, double a, double b); /* Function
to check if the particle dimensions are within the box */

int main()
{
      FILE *fptr;
      FILE *fptr2;

      /* Local variables */

      int K, P, Numfissions, Ninbox, Q;

      /* x0, y0, z0 is the position of the fission nucleus */
      /* x1, y1, z1, phi1, d1, costheta1 are positions of first neutron */
      /* x2, y2, z2, phi2, d2, costheta2 are positions of second neutron */
      double f, x0, y0, z0, phi1, phi2, d1, d2, costheta1, costheta2;

      double a, b, x1, y1, z1, x2, y2, z2;
      double pi;
      time_t t;

      pi = 3.142;
      P = 0;

      /* Select output file for error messages */
      fptr = fopen("chain1.err", "w");

      /* Initialize random number generator */
      srand((unsigned)time(&t));

      /* Ask the user for the number of fissions */
      printf("Enter number of fissions \n");
      scanf("%d", & Numfissions);

      /* Create results file  */
```

```
fptr2 = fopen("chain1.dat", "w");
if (fptr2 == NULL)
{
      fprintf(stderr, "Error writing to %s\n", "chain1.dat");
      fclose(stderr);
      return(1);
}

/* Start values for dimensions of box */
a = 2.0;
b = 0.1;

for (Q = 1;Q <= 20;Q++)
{
      Ninbox = 0;
      for (K = 1;K <= Numfissions;K++)
      {
            /* Find a random position within the box */
            /* for the nucleus */
            x0 = a * (randfunc(P) - 0.5);
            y0 = a * (randfunc(P) - 0.5);
            z0 = b * (randfunc(P) - 0.5);

           /* Find a random angles and distances for the 2 neutrons */
            phi1 = 2 * pi*randfunc(P);
            costheta1 = 2 * (randfunc(P) - 0.5);
            phi2 = 2 * pi*randfunc(P);
            costheta2 = 2 * (randfunc(P) - 0.5);
            d1 = randfunc(P);
            d2 = randfunc(P);

            /* Calculate the position of first neutron */
            x1 = x0 + d1 * sin(acos(costheta1))*cos(phi1);
            y1 = y0 + d1 * sin(acos(costheta1))*sin(phi1);
            z1 = z0 + d1 * costheta1;

            /* Calculate the position of second neutron */
            x2 = x0 + d2 * sin(acos(costheta2))*cos(phi2);
```

```
                y2 = y0 + d2 * sin(acos(costheta2))*sin(phi2);
                z2 = z0 + d2 * costheta2;

                /* Find out if first neutron is inside the box */
                if (checkin(x1, y1, z1, a, b) == 1)
                     Ninbox = Ninbox + 1;

                /* Find out if second neutron is inside the box */
                if (checkin(x2, y2, z2, a, b) == 1)
                     Ninbox = Ninbox + 1;

          }

          f = (double) Ninbox / (double) Numfissions;

          fprintf(fptr2, "%lf %lf\n", b / a, f);

          /* Make a smaller and b larger */
          /* We will show that a cube (a=b) */
          /* produces the best survival fraction */
          a = a - 0.1;
          b = b + 0.1;

     }
     fclose(fptr2);
     fclose(stderr);
}
double randfunc()
{
     /* Find a random number between 0 and 1 */
     double TOT;

     TOT = rand() % 1000;
     TOT = TOT / 1000;

     return TOT;

}

int checkin(double x, double y, double z, double a, double b)
```

```
{
        /* If the coordinates are within the box, return 1 */
        /* Otherwise return 0 */
        int I;

        if (x > -a / 2 && x < a / 2
              && y > -a / 2 && y < a / 2
              && z>-b / 2 && z < b / 2)
              I = 1;
        else
              I = 0;

        return I;

}
```

The graph of b/a against f is shown in Figure 13-5. This shows a peak for b/a = 1 or b = a. This would mean that the shape that gets closest to f = 1 is a cube.

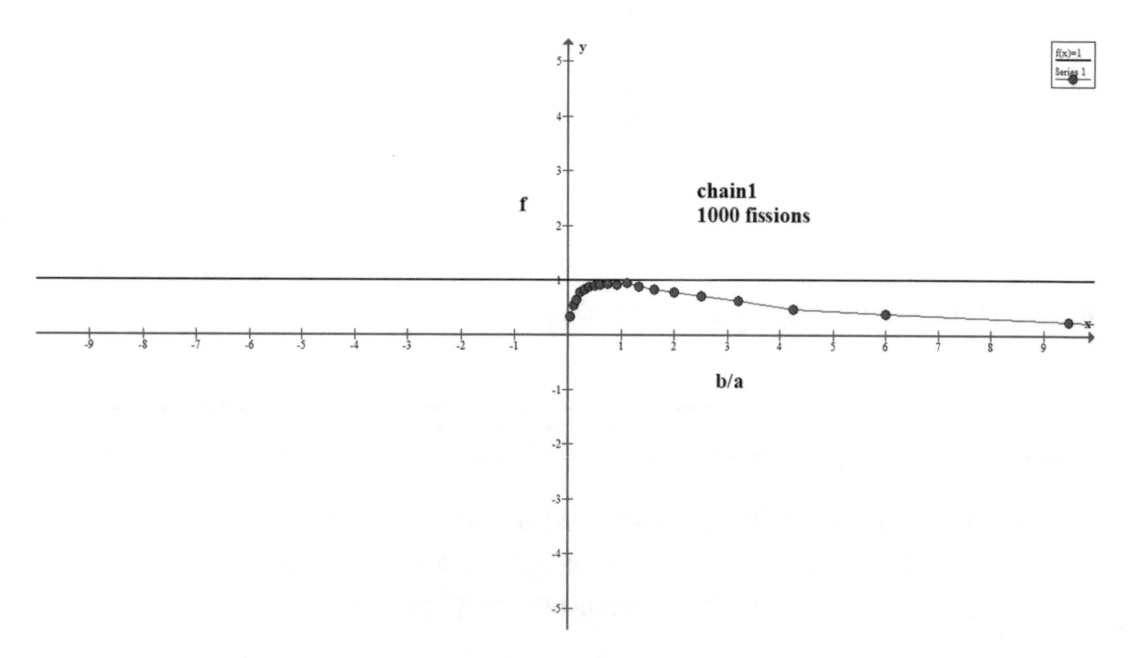

***Figure 13-5.** Dependence of survival fraction on shape*

Now we know that the cube is the best shape for our block, we can start with a block of small size and keep increasing the size. When we get to f = 1, then this is our point where our block reaches "critical mass."

This only requires a small change to the preceding program. At the start of the program, we can preset a and b both to 0.1 and increase their value by 0.1 at the end of our loop.

We can then plot a against f in our graph. The output from this is shown in Figure 13-6. Increasing the volume of the block increases our f value, as you would expect.

We have concentrated on rectangular blocks, but perhaps a spherical block would be the most efficient for creating a critical mass. This is given as an exercise at the end of the chapter.

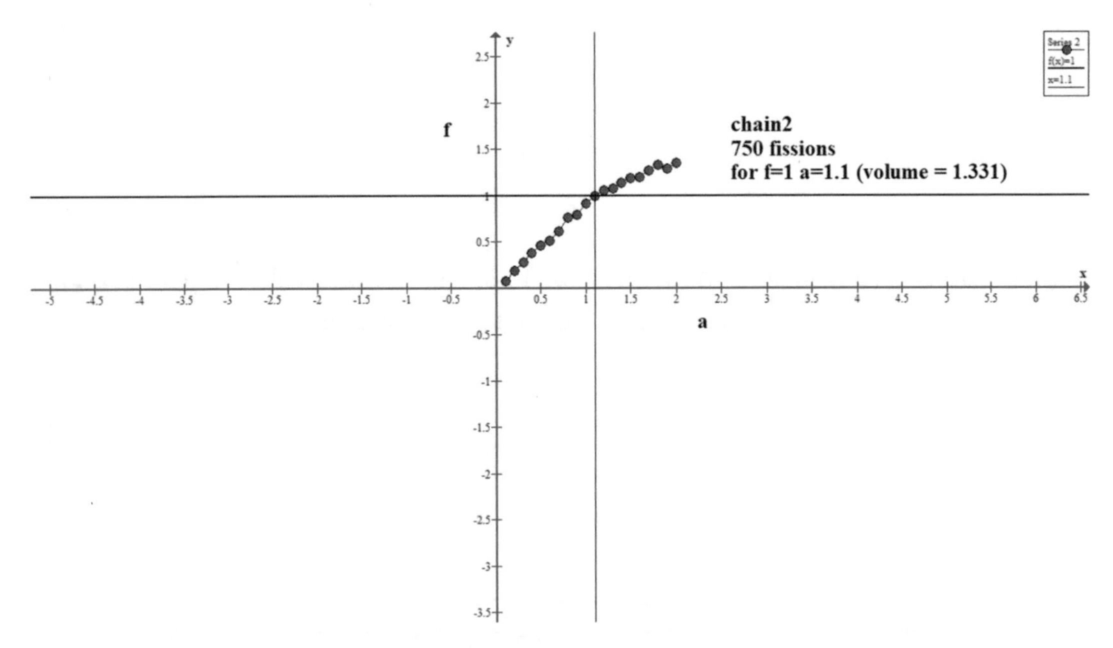

Figure 13-6. *Dependence of survival fraction on size*

EXERCISES

1. Amend the program in the chapter to find the survival fraction for a sphere. You can set an initial and final volume for the sphere and use the formula $V = (4/3) \pi r^3$ to get the corresponding values of r for the sphere.

   Try initial and final values for the volume to be

```
Vinit = 0.0001;
Vfin = 1.0;
```

APPENDIX

Answers to Problems

CHAPTER 1

1.

```
        /* ch1q1.c */
        /* Nested forloops */
#define _CRT_SECURE_NO_WARNINGS
#include <stdio.h>
#include <math.h>
#include <limits.h>
void main()
{
        int i,j,k,total; /* Store locations */
        total = 0;
        /* Loop i goes round 1000 times */
        /* Loop j goes round 1000 times */
        /* Loop k goes round 1000 times */
        for(i=0;i<1000;i++)
        {
              for(j=0;j<1000;j++)
              {
                    for(k=0;k<1000;k++)
                    {
                          total = total+1;
                    }
              }
        }
        /* Total count for 3 nested loops is 1000,000,000 */
        printf(" \n");
        printf("total is %d ", total);
```


P. Joyce, *Practical Numerical C Programming*, https://doi.org/10.1007/978-1-4842-6128-6

```
printf("The maximum value of INT = %d\n", INT_MAX);
       }
       2.
/*       ch1q2 */
/*       user enters points.*/
#define _CRT_SECURE_NO_WARNINGS
#include <stdio.h>
#include <math.h>
main()
{

            float total,average;
            float points[20] ;
            int i,numpoints;
            /* Enter the 20 numbers to be averaged */
            printf("\nEnter 20 numbers to be averaged");
            for(i=0;i<20;i++)
            {
                  scanf("%f",&points[i]);
}

            numpoints = 20;
       total = 0.0;

       /* Add up the numbers entered */
       for(i=0;i<numpoints;i++)
       {
            total =total + points[i];
       }
       /* Calculate the average and print out */
       printf("\ntotal is %f ", total);
       average = total/(float)numpoints;
       printf("\naverage is %f ", average);
}

3.
/* ch1q3 */
#define _CRT_SECURE_NO_WARNINGS
/* Function which returns an answer */
```

```
/* average of a set of numbers and returns it */

#include <stdio.h>
double getmarks(double pupils[]);
int main()
{
            double average;
            int i;
            float number;
            /* Array with marks for class is preset in the main part of the program */
            double marks[10]/* = { 10.6, 23.7, 67.9, 93.0, 64.2, 33.8 ,57.5 ,
            82.2 ,50.7 ,45.7 }*/;
            printf("\nEnter 10 numbers to be averaged");
            for(i=0;i<10;i++)
            {
                  scanf("%lf",&marks[i]);
            }
/* Call function getmarks. The function returns the max marks which is then
stored in pupil */
            number=0.0;
            average = getmarks(marks);
            printf("average is  = %f", average);
            return 0;
}
double getmarks(double marks[])
{
            int i;
            double average,total;
            total = 0.0;
            /* Go through all the pupils in turn and store the highest mark */
            for (i = 0; i < 10; ++i)
            {
                  total = total + marks[i];
            }
            average = total / 10;
            return average; /* Returns the average value to where the
            function was called */
}
```

4.

```
/* ch1q4 */
/* Create a file of company details */

#define _CRT_SECURE_NO_WARNINGS
#include<stdio.h>
        struct company {
             char name[13];
             int employees;
             float yearprofit;
        };
int main()
{
             int i, numread,nocomp;
             FILE *fp;
             struct company s1;

             fp = fopen("company.dat", "w");
             printf("\nEnter number of companies");
             scanf("\n%d",&nocomp);

             for (i = 0;i < nocomp;i++)
             {
                   printf("\nEnter company name (up to 13 characters)");
                   scanf("%s",s1.name);
                   printf("\nEnter number of employees");
                   scanf("%d",&s1.employees);
                   printf("\nEnter yearly profit");
                   scanf("%f",&s1.yearprofit);

                   fwrite(&s1, sizeof(s1), 1, fp);
             }

             fclose(fp);
             /* Reopen the file */
             fopen("company.dat", "r");

             /* Read and print out all of the records on the file */
```

```
            for (i = 0;i < nocomp/*0*/;i++)
            {
                  numread = fread(&s1, sizeof(s1), 1, fp);/* read into
                  structure s1 */
                  if (numread == 1)
                  {
                        /*printf( "Number of items read = %d ", numread );*/
                        /* Reference elements of structure by s1.company etc. */

                        printf("\ncompany Name : %s", s1.name);
                        printf("\nemployees : %d", s1.employees);
                        printf("\nyearly profit : %f", s1.yearprofit);
                  }
            else {
                  /* If an error occurred on read, then print out message */
                  if (feof(fp))
printf("Error reading company.dat : unexpected end of file fp is %p\n", fp);
                  else if (ferror(fp))
                  {
                        perror("Error reading company.dat");
                  }
            }
      }
      /* Close the file */
      fclose(fp);
      return;
}
```

5.

```
/* ch1q5 */
/* Reads and displays records from file which exceed the specified profit */
#define _CRT_SECURE_NO_WARNINGS
#include<stdio.h>
      struct company {
      char name[13];
      int employees;
      float yearprofit;
      };
```

```
int main()
{
        int i, numread,nocomp;
        float profittest;
        FILE *fp;
        struct company s1;

        printf("\nEnter number of companies on file");
        scanf("\n%d",&nocomp);

        printf("\nEnter profit to be exceeded");
             scanf("%f",&profittest);
        /* open the file */
        fp = fopen("company.dat", "r");

        /* Read and print out all of the relevant records on the file */
        for (i = 0;i < nocomp;i++)
        {
             numread = fread(&s1, sizeof(s1), 1, fp);/* Read into
             structure s1 */
             if (numread == 1)
             {
                    /*printf( "Number of items read = %d ", numread );*/
                    /* Reference elements of structure by s1.company etc. */
                    if(s1.yearprofit > profittest)
                    {
                           printf("\ncompany Name : %s", s1.name);
                           printf("\nemployees : %d", s1.employees);
                           printf("\nyearly profit : %f", s1.yearprofit);
                    }
             }
             else {
                    /* If an error occurred on read, then print out message */
                    if (feof(fp))

                           printf("Error reading company.dat : unexpected end of
                           file fp is %p\n", fp);
```

```
                else if (ferror(fp))
                      {
                            perror("Error reading company.dat");
                      }
                }
      }
      /* Close the file */
      fclose(fp);
      return;
}
```

6.

```
/* ch1q6 */
/* Reads and displays records from file which exceed the specified number of
employees  */
#define _CRT_SECURE_NO_WARNINGS
#include<stdio.h>
       struct company {
            char name[13];
            int employees;
            float yearprofit;
       };
int main()
{
            int i, numread,nocomp;
int emptest;
            FILE *fp;
            struct company s1;

            printf("\nEnter number of companies on file");
            scanf("\n%d",&nocomp);

            printf("\nEnter number of employees to be exceeded");
            scanf("%d",&emptest);
            /* Open the file */
            fp = fopen("company.dat", "r");

            /* Read and print out all of the relevant records on the file */
```

```
            for (i = 0;i < nocomp;i++)
            {
                  numread = fread(&s1, sizeof(s1), 1, fp);/* read into
                  structure s1 */
                  if (numread == 1)
                  {
                  /*printf( "Number of items read = %d ", numread );*/
                  /* Reference elements of structure by s1.company etc. */
                  if(s1.employees > emptest)
                  {
                        printf("\ncompany Name : %s", s1.name);
                        printf("\nemployees : %d", s1.employees);
                        printf("\nyearly profit : %f", s1.yearprofit);
                  }
            }
            else {
                  /* If an error occurred on read, then print out message */
                  if (feof(fp))

                        printf("Error reading company.dat : unexpected end of
                        file fp is %p\n", fp);

                  else if (ferror(fp))
                        {
                              perror("Error reading company.dat");
                        }
                  }
      }
      /* Close the file */
      fclose(fp);
      return;
}
```

7.

```
/* ch1q7.c */
/* Enter a switch value for specific function */
#define _CRT_SECURE_NO_WARNINGS
```

```
#include <stdio.h>

#include <math.h>
#define PI 3.14159265

int main()
{
        double angle, radianno, answer;
        int func;
        double  arccos, arcsin, arctan;
        double  expno, natlog, lb10;
        double  pownum, power, sqroot, fabsno;

    /* Prompt the user to enter the number which corresponds */
    /* to the function they want to find */

        printf("\nEnter number of which function you want");
        printf("\ncos = 1, sin = 2, tan = 3");
        printf("\narccos = 4, arcsin = 5, arctan = 6");
        printf("\nexp = 7, log = 8, log10 = 9");
        printf("\npow = 10, sqrt = 11, fabs = 12");

        printf("\n");
        scanf("%d",&func); /* Read in the number the user enters */

    /* Switch on that number */
    switch(func)
    {
    case 1:

        /* The cosine function */
        printf("cosine function:\n ");
        printf("Please enter angle in degrees:\n ");
        scanf("%lf", &angle);
        printf("You entered %lf\n", angle);
        radianno = angle * (2 * PI / 360);
        answer = cos(radianno);
        printf("cos of %lf is %lf\n", angle, answer);
        break;
```

```
case 2:
     /* The sine function */
     printf("sine function:\n ");
     printf("Please enter angle in degrees:\n ");
     scanf("%lf", &angle);
     printf("You entered %lf\n", angle);
     radianno = angle * (2 * PI / 360);
     answer = sin(radianno);
     printf("sin of %lf is %lf\n", angle, answer);
     break;
case 3:
     /* The tangent function */
     printf("tangent function:\n ");
     printf("Please enter angle in degrees:\n ");
     scanf("%lf", &angle);
     printf("You entered %lf\n", angle);
     radianno = angle * (2 * PI / 360);
     answer = tan(radianno);
     printf("tan of %lf is %lf\n", angle, answer);
     break;
case 4:
     /* The arccos function */
     printf("arccos function:\n ");
     printf("Please enter arccos:\n ");
     scanf("%lf", &arccos);
     printf("You entered %lf\n", arccos);
     radianno = acos(arccos);
     answer = radianno * (360 / (2 * PI));
     printf("arccos of %lf in degrees is %lf\n", arccos, answer);
     break;
case 5:
     /* The arcsin function */
     printf("arcsin function:\n ");
     printf("Please enter arcsin:\n ");
     scanf("%lf", &arcsin);
     printf("You entered %lf\n", arcsin);
```

```
        radianno = asin(arcsin);
        answer = radianno * (360 / (2 * PI));
        printf("arcsin of %lf in degrees is %lf\n", arcsin, answer);
        break;
    case 6:
        /* The arctan function */
        printf("arctan function:\n ");
        printf("Please enter arctan:\n ");
        scanf("%lf", &arctan);
        printf("You entered %lf\n", arctan);
        radianno = atan(arctan);
        answer = radianno * (360 / (2 * PI));
        printf("arctan of %lf in degrees is %lf\n", arctan, answer);
        break;

    /* Showing use of exp, log, and log10 functions */

    case 7:
        /* find exponent of entered number */
        printf("exponential function:\n ");
        printf("Please enter number:\n ");
        scanf("%lf", &expno);
        printf("You entered %lf\n", expno);

        answer = exp(expno);
        printf("exponent of %lf is %lf\n", expno, answer);
        break;
    case 8:
        /* Find natural logarithm of entered number */
        printf("natural logarithm function:\n ");
        printf("Please enter number:\n ");
        scanf("%lf", &natlog);
        printf("You entered %lf\n", natlog);
        answer = log(natlog);
        printf("natural logarithm of %lf is %lf\n", natlog, answer);
        break;
```

```
case 9:
     /* Find log to base 10 of entered number */
     printf("log to base 10 function:\n ");
     printf("Please enter number:\n ");
     scanf("%lf", &lb10);
     printf("You entered %lf\n", lb10);
     answer = log10(lb10);

     printf("log to base 10 of %lf is %lf\n", lb10, answer);
     break;

case 10:
     /* Showing use of pow, sqrt, and fabs functions */
     /* Find x raised to power y number */
     printf("power:\n ");
     printf("Please enter number:\n ");
     scanf("%lf", &pownum);
     printf("You entered %lf\n", pownum);
     printf("Please enter power:\n ");
     scanf("%lf", &power);
     printf("You entered %lf\n", power);

     answer = pow(pownum, power);
     printf("%lf raised to power %lf is %lf\n", pownum, power, answer);
     break;

case 11:
     /* Find square root of number */

     printf("square root:\n ");
     printf("Please enter number:\n ");
     scanf("%lf", &sqroot);
     printf("You entered %lf\n", sqroot);

     answer = sqrt(sqroot);
     printf("The square root of %lf is %lf\n", sqroot, answer);
     break;
```

```
        case 12:
             /* Find absolute value of number */
             printf("absolute value:\n ");
             printf("Please enter number:\n ");
             scanf("%lf", &fabsno);
             printf("You entered %lf\n", fabsno);

             answer = fabs(fabsno);
             printf("The absolute value of %lf is %lf\n", fabsno, answer);
             break;

        default:
             printf("\nError incorrect option entered");

        }
        return 0;
}
```

CHAPTER 2

1.

```
/*      regression */
/*      User enters points.*/
/*      Regression of x on y calculated */
#define _CRT_SECURE_NO_WARNINGS
#include <stdio.h>
#include <math.h>
void main()
{
        FILE *fp;

        double xpoints[12],ypoints[12];
        double sigmax,sigmay,sigmaxy,sigmaysquared,xbar,ybar;
        double fltcnt,sxy,syy,c,d;
        int i,points,invno;

        /* Clear x and y storage arrays to zero */
```

```
for(i=0;i<12;i++)
     {
           xpoints[i] = 0.0;
           ypoints[i] = 0.0;

     }

fp=fopen("regxony.dat","w");
/* User asked for number of points on scatter graph */
printf("enter number of points (max 12 ) \n");
scanf("%d", &points);
if(points>12)
{
     printf("error - max of 12 points\n");

}
else
{
     sigmax=0.0;
     sigmay=0.0;
     sigmaxy=0.0;

     sigmaysquared=0.0;

     /* User enters points */
     for(i=0;i<points;i++)
     {
           printf("enter point (x and y separated by space) \n");
           scanf("%lf %lf", &xpoints[i], &ypoints[i]);
           sigmax=sigmax+xpoints[i];
           sigmay=sigmay+ypoints[i];
           sigmaxy=sigmaxy+xpoints[i]*ypoints[i];

           sigmaysquared=sigmaysquared+pow(ypoints[i],2);
     }
     printf("points are \n");
     for(i=0;i<points;i++)
     {
           printf(" \n");
           printf("%lf %lf", xpoints[i], ypoints[i]);
           fprintf(fp,"%lf\t%lf\n",xpoints[i], ypoints[i]);
     }
```

```
        printf(" \n");
        fltcnt=(double)points; /* Set fltcnt to value of points (as double) */

        /* Regression variables calculated */
        xbar=sigmax/fltcnt;
        ybar=sigmay/fltcnt;
        sxy=(1/fltcnt)*sigmaxy-xbar*ybar;

        syy=(1/fltcnt)*sigmaysquared-ybar*ybar;

        d=sxy/syy;
        c=xbar-d*ybar;

        /* Regression line */
        printf("Equation of regression line x on y  is\n ");
        printf(" x=%lf + %lfy", c,d);
    }
    fclose(fp);
    return;

}
```

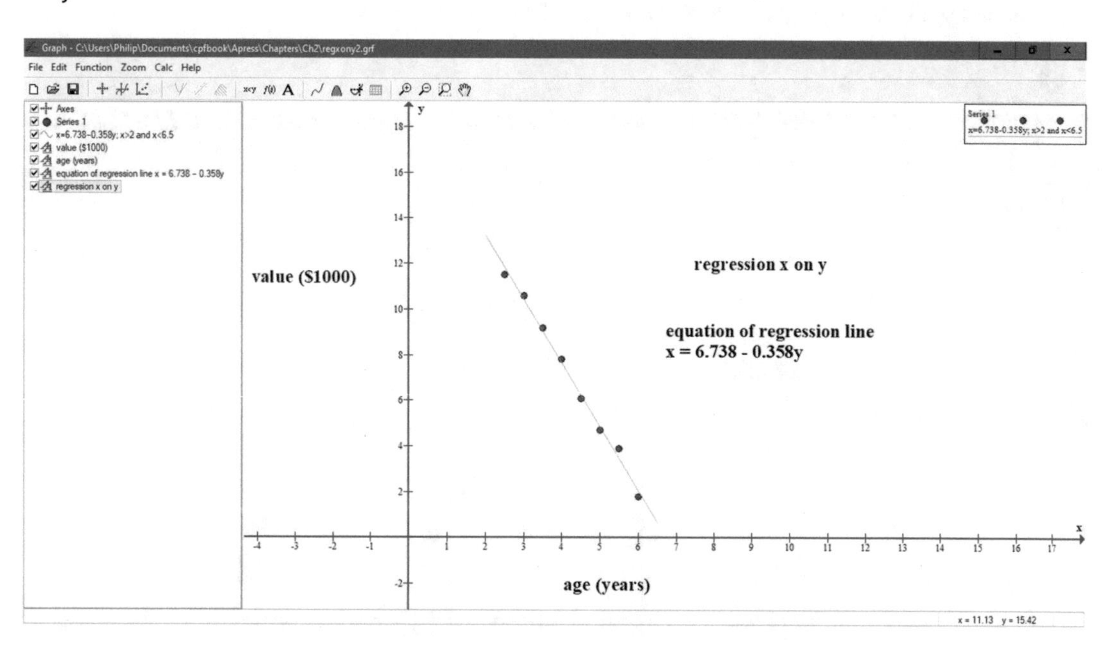

2.

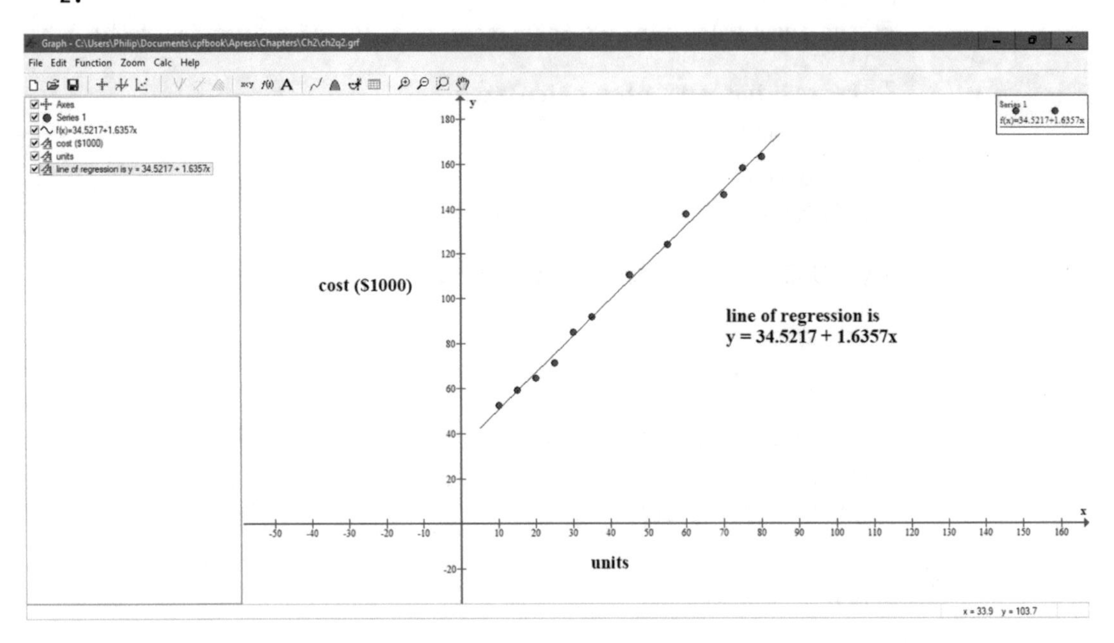

50 units would be between 110 and 120 from the graph, or if you substitute x=50 into the equation of the line of regression, you should get 116.3067.

3.

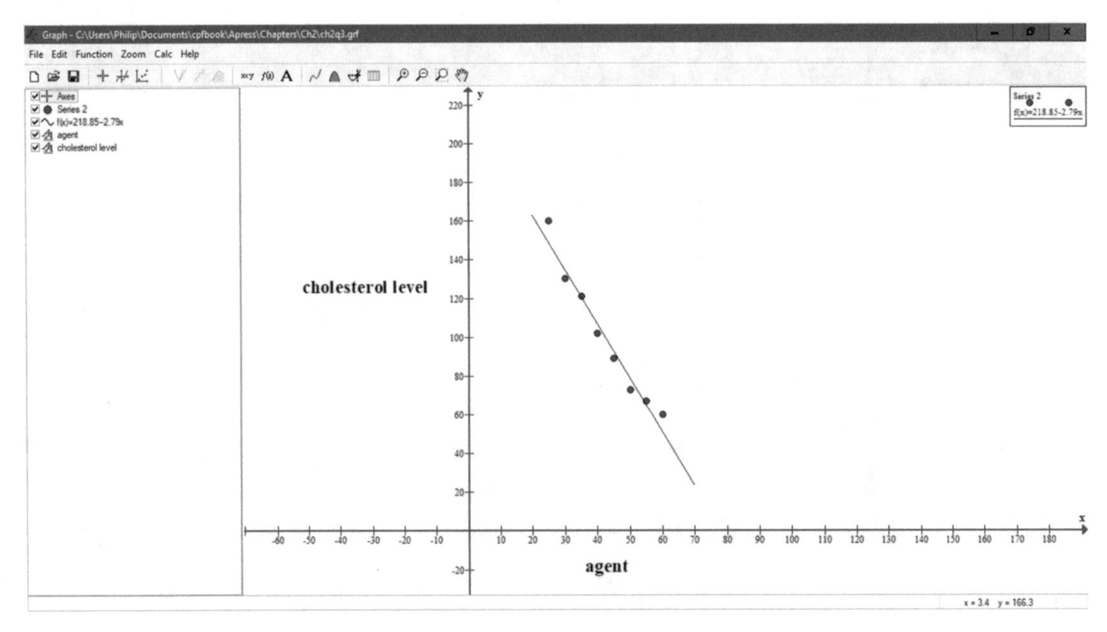

The cholesterol level drops by increasing the use of the chemical agent.

CHAPTER 3

1. The data is shown in the following.

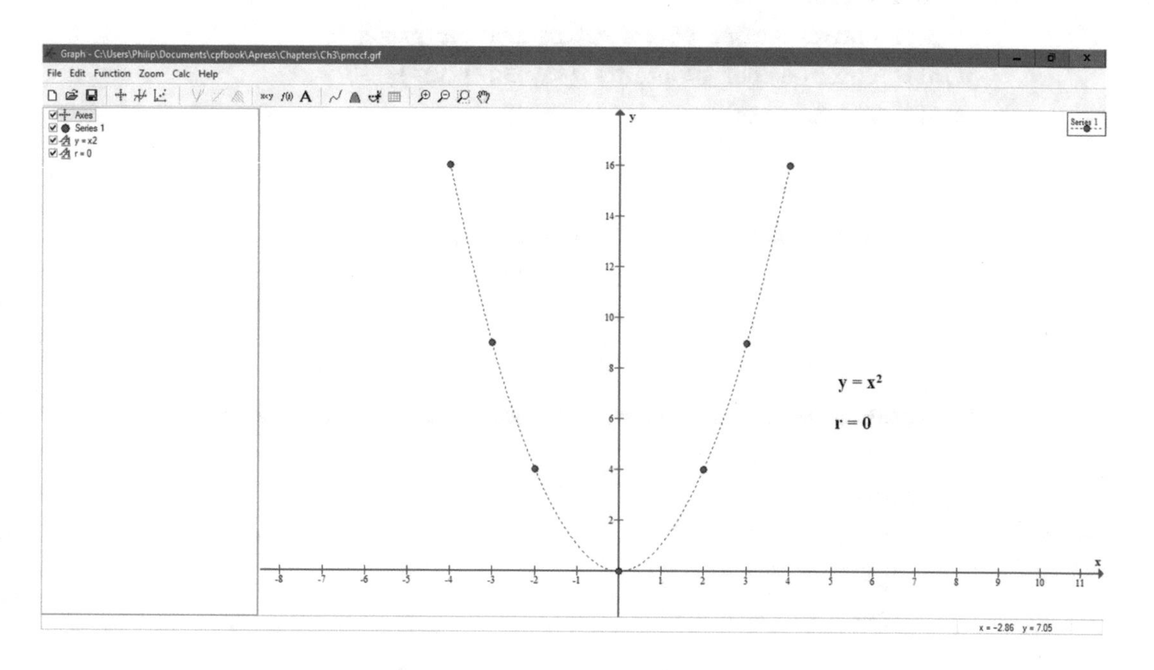

The data points for the curve y = x^2.

So you can see there is correlation but this is not linear correlation so we get a PMCC of zero.

The code for this is as follows:

```
/*Product moment correlation coefficient */
#define _CRT_SECURE_NO_WARNINGS
#include <stdio.h>
#include <math.h>
main()
{

       double xpoints[10], ypoints[10];

/* Variables are named as in the formulas used in the text of the chapter */
/* e.g., Σx is called "sigmax" */

       double sigmax, sigmay, sigmaxsquared, sigmaysquared, xbar, ybar, sigmaxy;
       double sxy, sxx, syy, sx, sy, r;
```

```
int i, points; /* User-entered number of scatter graph points */
double fltcnt; /* Number of points as a double variable */
FILE *fp;
fp=fopen("pmccf.dat","w");
/* User enters number of points in scatter graph */
printf("enter number of points (max 10 ) \n");
scanf("%d", &points);
if (points > 10)
{
     printf("error - max of 10 points\n");

}
else
{
/* Variables used for summing data values cleared to zero */
     sigmax = 0.0;
     sigmay = 0.0;
     sigmaxy = 0.0;
     sigmaxsquared = 0.0;
     sigmaysquared = 0.0;

     /* User enters points in scatter graph */
     for (i = 0;i < points;i++)
     {
            printf("enter point (x and y separated by space) \n");
            scanf("%lf %lf", &xpoints[i], &ypoints[i]);
            /* totals incremented by x and y points */
            sigmax = sigmax + xpoints[i];
            sigmay = sigmay + ypoints[i];
            sigmaxy = sigmaxy + xpoints[i] * ypoints[i];
            sigmaxsquared = sigmaxsquared + pow(xpoints[i], 2);
            sigmaysquared = sigmaysquared + pow(ypoints[i], 2);
     }
     printf("points are \n");
     for (i = 0;i < points;i++)
     {
            printf(" \n");
```

```
                printf("%lf %lf", xpoints[i], ypoints[i]);
                fprintf(fp,"%lf\t%lf\n",xpoints[i], ypoints[i]);
            }
            printf(" \n");

            /* Convert number of points as a double variable */
            /* for use in the formulas. Store this in variable fltcnt */
            fltcnt = (double)points;
            /* variables in PMCC formula calculated */
            xbar = sigmax / fltcnt;
            ybar = sigmay / fltcnt;

            syy = (1 / fltcnt)*sigmaysquared - ybar * ybar;

            sxx = (1 / fltcnt)*sigmaxsquared - xbar * xbar;
            sx = sqrt(sxx);
            sy = sqrt(syy);
            sxy = (1 / fltcnt)*sigmaxy - xbar * ybar;

            /* PMCC value calculated */
            r = sxy / (sx*sy);
            printf("r is %lf", r);
        }

        fclose(fp);
}
```

CHAPTER 4

1. Your code should be similar to the following code:

```
/* ch4q1.c */
/* Stock price predictor simulation */
/* from Day values */

#define _CRT_SECURE_NO_WARNINGS
#include <stdio.h>
#include <math.h>
#include <stdlib.h>
```

```
#include <time.h>

double calcrand();/* Function to calculate our random value for x using the
formula */
void avstdvar(double dayvals[]);/* Function to calculate variance &
standard dev */

FILE *fp;

        double c,c0,c1,c2,d1,d2,d3;
        double q,d,F,y,t;

        int i,j;
        time_t tim;

        int  n;
        double average, variance, std_deviation, sum = 0.0, sum1 = 0.0;

/* The PDR prefix to variables denotes Periodic Day Return */
        double PDRaverage,PDRvariance,PDRstd_deviation,pdrsum,nextval,lastval,
        drift,epsilon,exptest,nitest;
        double pdr[50];

void main()
{
        FILE *fp;
        /* Array containing day stock prices starting with yesterday and
        moving backward */

        /* The part of the array following these preset values will contain our /*
        /* calculated stock price values. So the whole array can be printed out */
        /* on our graph */

        double dayvals[50];

        double value,testval;
        int j;

        fp=fopen("ch4q1.dat","w");
```

```
srand((unsigned) time(&tim)); /* Set up random number function */

for(i=0;i<50;i++)
{
     /* Clear predicted rate array */
     pdr[i]=0.0;
}
for(i=19;i<50;i++)
{
     /* Clear the end part of our values array for our predicted vales */
     dayvals[i]=0.0;

}

/* Enter number of predicted daily rates (PDR) */

printf("enter number of days up to 50\n");
scanf("%d",&n);

/* Enter data one element at a time*/

for(j=0;j<n;j++)
{
     printf("day %d ",j+1);

     printf("enter price\n");
     scanf("%lf",&value);
     dayvals[j] = value;

}
j=0;
/* Write historical rates to output file */

for(i=18;i>-1;i--)
{
     fprintf(fp,"%d\t%lf\n",j,dayvals[i]);
     j++;
}
```

```
    /* Calc PDRs - if you enter n days, there will be n-1 PDRs */
    for(j=0;j<n-1;j++)
    {
        pdr[j]=log(dayvals[j]/dayvals[j+1]);

        printf("pdr[j] = %lf\n",pdr[j]);
    }
    /* Compute the sum of all elements */
    pdrsum=0.0;
    for (i = 0; i < n-1; i++)

    {

        pdrsum = pdrsum + pdr[i];
    }

    PDRaverage = pdrsum / (double)(n-1);
    /* Call function to calculate statistical values */

    avstdvar(dayvals);
    /* Calculate drift */

    drift=PDRaverage-(PDRvariance/2);
    printf("drift is %lf\n",drift);
    printf("PDRaverage is %lf\n",PDRaverage);

    lastval=dayvals[0];

    /* Calculate values using formula from the chapter*/
/*Today's Stock Price = Yesterday's Stock price * exp( Drift + Random Change)*/
/* We use the variable nextval for Today's Stock Price */
/* and the variable lastval for = Yesterday's Stock price */

/* and PDRstd_deviation*calcrand() for Random Change */
    /* nextval=lastval*exp(drift+PDRstd_deviation*calcrand()) */
    for (i = 19; i < 38; i++)

    {
        nitest=calcrand(); /* get random number using algorithm */
        exptest=exp(drift+PDRstd_deviation*nitest); /* get exp part of
        formula */
```

```
            nextval=lastval*exptest; /* lastval*exp part as above */
            printf("exptest is %lf\n",exptest);

            printf("nitest is %lf\n",nitest);
            printf("nextval is %lf\n",nextval);

            fprintf(fp,"%d\t%lf\n",i,nextval);

            lastval=nextval; /* set last value for the next iteration */

      }
      fclose(fp);
      return;
}
double calcrand()
{
      /* Set values for cumulative normal distribution formula */
      c0=2.515517;
      c1=0.802853;
      c2=0.010328;
      d1=1.432788;
      d2=0.189269;

      d3=0.001308;
      y=rand()%1000;/* Generate random number between 0 and 1 */

      y=y/1000;
      if(y>=0.5)
            q=1-y;
      else

            q=y;
      /* Apply the Cumulative Normal Distribution Algorithm */
      t=sqrt(log(1/pow(q,2)));
      c=c0+c1*t+c2*pow(t,2);

      d=1+d1*t+d2*pow(t,2)+d3*pow(t,3);

      F=t-(c/d);
      /* Use the symmetry of the Cumulative Normal Distribution graph */
      if(y < 0.5)
```

```
        {
             y=-1.0*F;
        }
        else if(y == 0.5)
             y=0;
        else
             y=F;
        printf("y  = %lf\n",y);
        return y;
}
void avstdvar(double dayvals[])
{
        /* Average, standard deviation, variance processing */
        sum = 0.0;
        sum1 = 0.0;
        /* Compute the sum of all elements */
        for (i = 0; i < n; i++)
        {
             sum = sum + dayvals[i];
        }
        average = sum / (double)n;
        /* Compute variance and standard deviation */
        for (i = 0; i < n; i++)
        {
             sum1 = sum1 + pow((dayvals[i] - average), 2);
        }
        variance = sum1 / (double)n;
        /* Compute PDRvariance and PDRstandard deviation */
        sum1=0.0;
        for (i = 0; i < n-1; i++)
        {
             sum1 = sum1 + pow((pdr[i] - PDRaverage), 2);
        }
```

```
        PDRvariance = sum1 / (double)(n-1);
        std_deviation = sqrt(variance);
        PDRstd_deviation = sqrt(PDRvariance);
        printf("Average of all elements = %lf\n", average);
        printf("variance of all elements = %lf\n", variance);
        printf("Standard deviation = %lf\n", std_deviation);
        printf("PDRvariance of all elements = %lf\n", PDRvariance);

        printf("PDRStandard deviation = %lf\n", PDRstd_deviation);

}
```

The graph generated by this code is shown in the following.

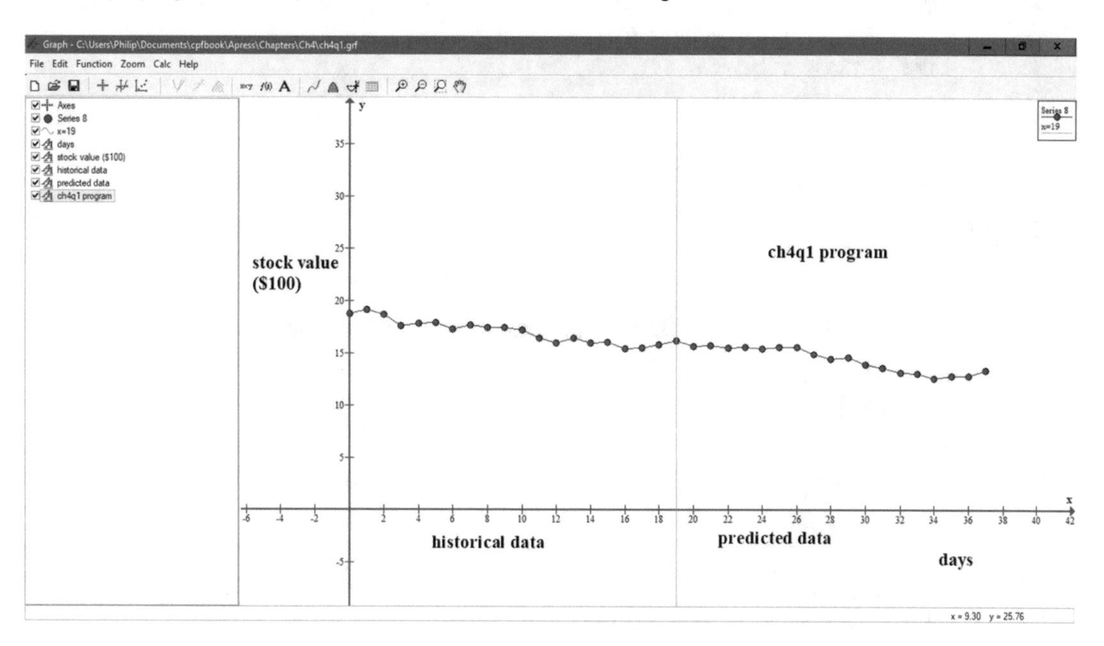

This graph differs from the one in the chapter as the trend goes down in the historical data and correspondingly so in the predicted data. The predicted data part of your graph may differ from this one as the random numbers you will have generated will have differed from the ones when this graph was generated.

2. The program should be similar to this

```
/* assetq2.c */
/* Stock price predictor simulation */
/* produces cumulative normal */
/* distribution function graph */
```

```
/* using inverse function technique */

#define _CRT_SECURE_NO_WARNINGS
#include <stdio.h>
#include <math.h>
#include <stdlib.h>

#include <time.h>

FILE *fp;
        double c,c0,c1,c2,d1,d2,d3;

        double q,d,x,y,t;
        int i;

        time_t tim;

        int  n;

        double probvalue;
int main()

{
        /* Set values for cumulative normal distribution formula */
        c0=2.515517;
        c1=0.802853;
        c2=0.010328;
        d1=1.432788;
        d2=0.189269;

        d3=0.001308;
        fp=fopen("assetq2.dat","w");

        srand((unsigned) time(&tim)); /* Set up random number function */
        for(i=0;i<50;i++)

        {
             y=rand()%1000; /* Generate random number between 0 and 1 */
             y=y/1000;

             probvalue=y; /* Store the current probability value */
             /* Use the symmetry of the Cumulative Normal Distribution graph */
```

```
    if(y>=0.5)
            q=1-y;
        else

            q=y;
        /* Calculate the values in the formula */
        t=sqrt(log(1/pow(q,2)));
        c=c0+c1*t+c2*pow(t,2);

        d=1+d1*t+d2*pow(t,2)+d3*pow(t,3);

        x=t-(c/d);
        /* Use the symmetry of the Cumulative Normal Distribution graph */
        if(y < 0.5)

        {

            y=-1.0*x;
        }
        else if(y == 0.5)
            y=0;
        else

            y=x;
        /* Store the calculated x value and the current*/

        /* probability value in the file */
        fprintf(fp,"%lf\t%lf\n",y,probvalue);

    }

    fclose(fp);
    return;

}
```

This is an example of the graph produced by this code. It shows a good Cumulative Normal Distribution showing the accuracy of the algorithm method for finding the x values.

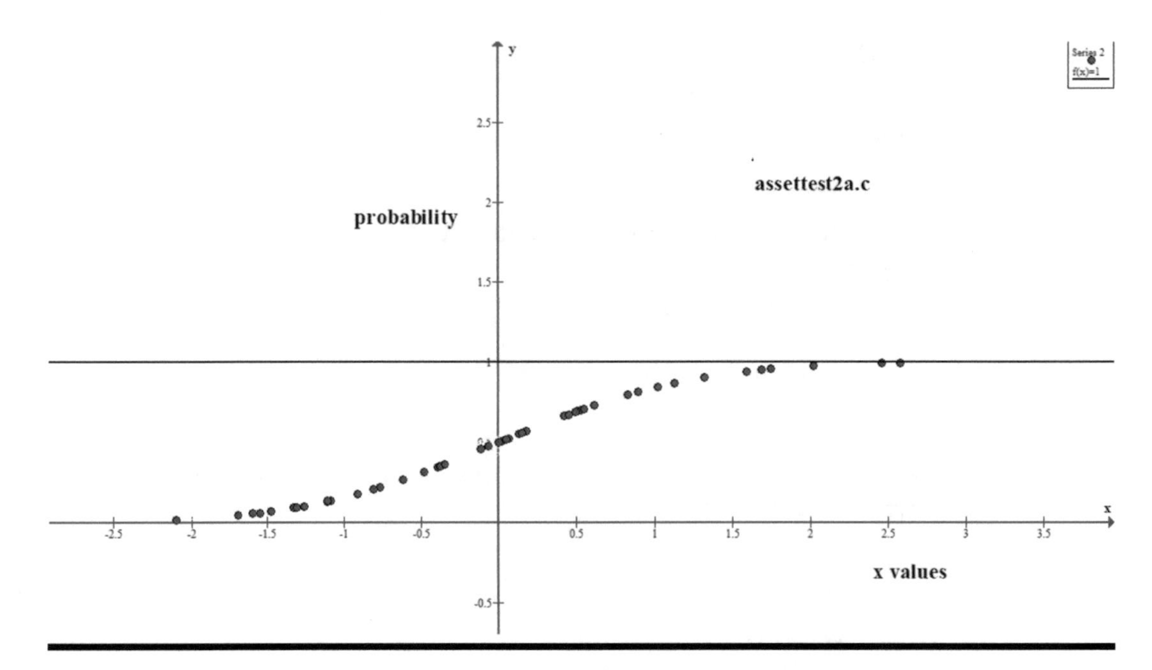

CHAPTER 5

1. This is an example of what your program should look like.

```
/* mtest.c */
/* Supermarket reordering simulation */
/* Print out the superm.dat file */
#define _CRT_SECURE_NO_WARNINGS
#include <stdio.h>
#include <math.h>
#include <stdlib.h>

#include <time.h>
struct super {
   int ID;
   char desc[11];
   int limit;
   int numstock;
   char address[30];

};
struct super s1;
```

```
        struct super s2;
void main()
{
        FILE *fp;
         int numread, i;
        /* Open supermarket file */
        fp = fopen("superm.dat", "r");
        /* Read and print out all of the records on the file */
        printf("\nID DESCRIPTION LIMIT NUMBER IN STOCK ADDRESS");
        for(i=0;i<17;i++)
        {
             numread=fread(&s2, sizeof(s2), 1, fp);
             if(numread == 1)
             {
                   printf("\n%2d : %s : %d : %d : %s", s2.ID,s2.desc,s2.limit,
                   s2.numstock,s2.address); /* Note the 2d as we want
                   2 digits */
             }
             else {
                   /* If an error occurred on read, then print out message */
                         if (feof(fp))
                            printf("Error reading superm.dat : unexpected end
                            of file fp is %p\n",fp);
                         else if (ferror(fp))
                   {
                            perror("Error reading superm.dat");
                         }
             }
        }
        /* Close the file */
        fclose(fp);
        return;
}
```

The output to the command line should look like this:

```
ID      DESCRIPTION      LIMIT    NUMBER IN STOCK      ADDRESS
 4      : Brie           : 23     : 50                 : 95,West Park St
 7      : Gouda          : 34     : 51                 :  2,North Park St
 9      : Edam           : 44     : 52                 : 17,New Gate St
11      : Camembert      : 25     : 53                 : 12,Toll Av
14      : Cheshire       : 34     : 66                 :  5,State Rd
16      : Cheddar        : 51     : 30                 : 63,Madison St
17      : Pecorino       : 23     : 56                 : 12,East Park St
19      : Manchego       : 44     : 57                 : 14,May St
23      : Provolone      : 35     : 58                 : 20,Oregon Way
24      : Parmigiano     : 40     : 59                 : 10,Park St
27      : Mascarpone     : 40     : 60                 : 31,Queen St
31      : Mozzarella     : 42     : 61                 : 19,Hope Av
32      : Feta           : 45     : 62                 : 13,Charles Av
35      : Gruyere        : 47     : 63                 : 54,Tower St
38      : Monterey       : 41     : 63                 : 11,Cardew Av
44      : Gorgonzola     : 54     : 68                 : 26,Jones St
47      : Stilton        : 58     : 69                 : 57,Lower St
```

CHAPTER 6

1. Your program should be similar to this.

```
/* fltcnt.c */
/* displays count of flights */
/* in the flightcnt.dat file */
#define _CRT_SECURE_NO_WARNINGS
#include <stdio.h>
#include <math.h>
#include <stdlib.h>
#include <time.h>

#include <string.h>
void main()
{
        FILE *fltcnt;
```

```
struct flightcount {
       int count;
};

struct flightcount fc;
       fltcnt = fopen("flightcnt.dat","r");
       fread(&fc, sizeof(fc), 1, fltcnt);

       printf(" Number of flights : %d", fc.count);

       fclose(fltcnt);

}
```

2. Your program will be similar to this.

```
/* createdep.c */
/* Creates departures file */

/* Prints out the records sequentially */
#define _CRT_SECURE_NO_WARNINGS
#include<stdio.h>

#include <string.h>
struct departures {
       char posn[3];
       char flight_no[8];
       char sch_departure_time[6];
       char exp_departure_time[6];
       char destination[15];
       char checkingate[5];
       char remarks[14];

};
struct depcount {
       int count;
};
void main()
{

       int i,numread;
       FILE *fpdep;
       FILE *fltcnt;
```

```
struct departures s1;
struct depcount fc={17};

struct depcount fcr;
struct departures s10 = {"1","AA1232","07:00","07:00","CHICAGO","A4",""};
struct departures s11 = {"2","BA123","07:05","07:05","LONDON","A1",""};
struct departures s12 = {"3","AA4517","07:08","07:15","BOSTON","B4",""};
struct departures s13 = {"4","AF123","07:10","07:10","PARIS","C2",""};
struct departures s14 = {"5","NH444","07:20","07:20","TOKYO","A7",""};
struct departures s15 = {"6","DJ144","07:22","07:22","MUMBAI","D4",""};
struct departures s16 = {"7","AZ2348","07:23","07:25","WASHINGTON","B1",""};
struct departures s17 = {"8","VS9745","07:25","07:26","TORONTO","A3",""};
struct departures s18 = {"9","DL5816","07:30","07:30","CHICAGO","D9",""};
struct departures s19 = {"10","KL5393","07:33","07:33","MANCHESTER","C3",""};
struct departures s20 = {"11","AZ4627","07:35","07:40","ROME","C7",""};
struct departures s21 = {"12","VS4677","07:40","07:40","NEW ORLEANS","B3",""};
struct departures s22 = {"13","SQ125","07:45","07:45","FRANKFURT","E6",""};
struct departures s23 = {"14","EI5666","07:48","07:48","LONDON","E4",""};
struct departures s24 = {"15","WS2321","07:50","07:50","DULLES","E1",""};
struct departures s25 = {"16","AA197","07:55","08:00","SAN FRANCISCO","E3",""};

struct departures s26 = {"17","B57321","07:58","07:48","SARASOTA","E2",""};
/* Create the file depcnt.dat which will contain */
/* the current number of flights in departures.dat. */
/* This file can then be updated when flights are */

/* removed from departures.dat to keep a running total */
fltcnt = fopen("depcnt.dat","w");
fwrite(&fc, sizeof(fc), 1, fltcnt);

fclose(fltcnt);
fltcnt = fopen("depcnt.dat","r");
fread(&fcr, sizeof(fcr), 1, fltcnt);
printf(" Number of flights : %d", fcr.count);
fclose(fltcnt);

/* Open the departures file */

   fpdep = fopen("departures.dat", "w");

/* Write details of each flight to file*/
```

```
/* From the structures defined earlier */
fwrite(&s10, sizeof(s1), 1, fpdep);
fwrite(&s11, sizeof(s1), 1, fpdep);
fwrite(&s12, sizeof(s1), 1, fpdep);
fwrite(&s13, sizeof(s1), 1, fpdep);
fwrite(&s14, sizeof(s1), 1, fpdep);
fwrite(&s15, sizeof(s1), 1, fpdep);
fwrite(&s16, sizeof(s1), 1, fpdep);
fwrite(&s17, sizeof(s1), 1, fpdep);
fwrite(&s18, sizeof(s1), 1, fpdep);
fwrite(&s19, sizeof(s1), 1, fpdep);
fwrite(&s20, sizeof(s1), 1, fpdep);
fwrite(&s21, sizeof(s1), 1, fpdep);
fwrite(&s22, sizeof(s1), 1, fpdep);
fwrite(&s23, sizeof(s1), 1, fpdep);

fwrite(&s24, sizeof(s1), 1, fpdep);
fwrite(&s25, sizeof(s1), 1, fpdep);

fwrite(&s26, sizeof(s1), 1, fpdep);

/* Close the file */

   fclose(fpdep);

/* Reopen the file */

fopen("departures.dat", "r");
/* Read and print out all of the records on the file */
printf("\n Flight :Sched: Exp: Gate Destination Remarks");
for(i=0;i<17;i++)

{

     numread=fread(&s1, sizeof(s1), 1, fpdep);
     if(numread == 1)
     {
          printf("\n :%s\t%s\t%s\t%s\t%s\t%s\t%s", s1.posn,s1.
          flight_no,s1.sch_departure_time,s1.exp_departure_time,s1.
          checkingate,s1.destination,s1.remarks);
     }
```

```
          else {

                 /* If an error occurred on read, then print out message */

                        if (feof(fpdep))

                           printf("Error reading departures.dat : unexpected
                           end of file fpdep is %p\n",fpdep);
                        else if (ferror(fpdep))
                  {
                           perror("Error reading departures.dat");
                        }

          }
     }

     /* Close the file */

     fclose(fpdep);

}
```

CHAPTER 7

1. Your program should be similar to this.

```
/* createplant.c  */
/* Reads from file */
/* Prints out the records sequentially */

/* Finds specific records and prints them */

#define _CRT_SECURE_NO_WARNINGS
#include<stdio.h>

struct fpress {
   int ID;  /* ID for the device */
   float llimit;  /* Lower limit for pressure */
   float press;  /* Current pressure */

   float ulimit;  /* Upper limit for pressure */

};
```

```
int main()
{
        int i,numread;
        FILE *fp;
        struct fpress s1;
        struct fpress s2;

        /* Preset structures with values for */
        /* the elements in the structure fpress */

        struct fpress s10 = {4,10.0,23.0,50.0};
        struct fpress s11 = {7,11.0,34.0,51.0};
        struct fpress s12 = {9,12.0,44.0,52.0};
        struct fpress s13 = {11,13.0,25.0,53.0};
        struct fpress s14 = {14,14.0,34.0,54.0};
        struct fpress s15 = {16,15.0,51.0,55.0};
        struct fpress s16 = {17,16.0,23.0,56.0};
        struct fpress s17 = {19,17.0,44.0,57.0};
        struct fpress s18 = {23,18.0,35.0,58.0};
        struct fpress s19 = {24,19.0,40.0,59.0};
        struct fpress s20 = {27,20.0,40.0,60.0};
        struct fpress s21 = {31,21.0,42.0,61.0};
        struct fpress s22 = {32,22.0,45.0,62.0};
        struct fpress s23 = {35,23.0,47.0,63.0};
        struct fpress s24 = {38,24.0,41.0,63.0};

        struct fpress s28 = {44,28.0,54.0,68.0};
        struct fpress s29 = {47,29.0,58.0,69.0};

        /* Open the pressure file */

        fp = fopen("pressure.bin", "w");

        /* Write details of each structure to file*/
        /* From the structures defined earlier */

        fwrite(&s10, sizeof(s1), 1, fp);
        fwrite(&s11, sizeof(s1), 1, fp);
        fwrite(&s12, sizeof(s1), 1, fp);
        fwrite(&s13, sizeof(s1), 1, fp);
        fwrite(&s14, sizeof(s1), 1, fp);
```

```
fwrite(&s15, sizeof(s1), 1, fp);
fwrite(&s16, sizeof(s1), 1, fp);
fwrite(&s17, sizeof(s1), 1, fp);
fwrite(&s18, sizeof(s1), 1, fp);
fwrite(&s19, sizeof(s1), 1, fp);
fwrite(&s20, sizeof(s1), 1, fp);
fwrite(&s21, sizeof(s1), 1, fp);
fwrite(&s22, sizeof(s1), 1, fp);
fwrite(&s23, sizeof(s1), 1, fp);
fwrite(&s24, sizeof(s1), 1, fp);
fwrite(&s28, sizeof(s1), 1, fp);
fwrite(&s29, sizeof(s1), 1, fp);

/* Close the file */

fclose(fp);

/* Reopen the file */

fp=fopen("pressure.bin", "r");

/* Read and print out all of the records on the file */

for(i=0;i<17;i++)

{
      numread=fread(&s2, sizeof(s2), 1, fp);
      if(numread == 1)

      {

           printf("\nID : %d lower limit : %f pressure : %f upper
           limit : %f", s2.ID,s2.llimit,s2.press,s2.ulimit);
     }
     else {

           /* If an error occurred on read, then print out message */

                 if (feof(fp))

                    printf("Error reading pressure.bin : unexpected
                    end of file fp is %p\n",fp);
```

```
                    else if (ferror(fp))
                     {
                       perror("Error reading pressure.bin");
                    }

            }
        }

        /* Close the file */

        fclose(fp);

}
```

2. Your program should be similar to this.

```
/* plant.c */
/* Industrial plant simulation */
#define _CRT_SECURE_NO_WARNINGS
#include <stdio.h>
#include <math.h>
#include <stdlib.h>

#include <time.h>
struct fpress {
   int ID; /* ID for the device */
   float llimit; /* Lower limit for pressure */
   float press; /* Current pressure */
   float ulimit; /* Upper limit for pressure */

};
int main()
{

       FILE *fp;

       struct fpress s2;
       struct fpress st[17];
       int i;
       int IDtoamend;
       int IDfound;

       float fppress;

       /* Open pressure file */
```

```
fp = fopen("pressure.bin", "r");
for (i = 0;i < 17;i++)
{
     /* Read each pressure data from file sequentially */
     fread(&s2, sizeof(s2), 1, fp);

     /* Print pressure data each component */
     st[i].ID = s2.ID;
     st[i].llimit = s2.llimit;
     st[i].press = s2.press;
     st[i].ulimit = s2.ulimit;

     printf("\nID : %d lower limit : %f pressure : %f upper limit : %f",
     s2.ID,s2.llimit,s2.press,s2.ulimit);

}

fclose(fp);

/* Ask user to enter the ID */
/* If the ID is not in the file, */
/* the user is prompted to enter */

/* it again. */

IDfound=0;
do {
     fp = fopen("pressure.bin", "r+");
     /* Ask user to enter ID */
     printf("\nenter ID   \n");

     scanf("%d", &IDtoamend);
     printf("\n ID is %d",IDtoamend);
     for (i = 0;i < 17;i++)
     {
          fread(&s2, sizeof(s2), 1, fp);

          if(IDtoamend == s2.ID)
          {
               /* Valid ID found */
               IDfound=1;
               break;
```

```
            }

        }
        if(IDfound==0)

            printf("\nID not found");
        fclose(fp);

    } while( IDfound==0);

    printf("\n ID is %d",IDtoamend);

    /* User is prompted to enter the current pressure */
    printf("\nenter current pressure   \n");

    scanf("%f", &fppress);

    printf("\n current pressure is %f",fppress);
        for (i = 0;i < 17;i++)

        {

            if(IDtoamend == st[i].ID)
            {
                /* Test if the current pressure is */
                /* below the lower limit or above the */
                /* upper limit. If either is true, then */
                /* an alert message is output */

                printf("\n struct lower press is %f",st[i].llimit);
                printf("\n struct upper press is %f",st[i].ulimit);
                if(fppress < st[i].llimit)
                    printf("\n ALERT! Pressure is below lower limit");
                if(fppress > st[i].ulimit)
                    printf("\n ALERT! Pressure is above upper limit");

            }

        }

    }
```

3. Your program should be similar to this.

```
/* plantbam.c */
/* Industrial plant simulation */
/* Power plant temperature and flow rate */
/* Allows amendments to tempflow.bin file */
/* Tests if the amendment is above the hightemp */
/* value and outputs an alert if it is */

#define _CRT_SECURE_NO_WARNINGS
#include <stdio.h>
#include <math.h>
#include <stdlib.h>
#include <time.h>

/* Structure definition for each device on file */
struct fplant {
   int ID; /* ID of device */
   float temp; /* Current temperature of device */
   float flowrate; /* Current flow rate of device */
   float hightemp; /* Maximum temperature of device */
   float highflow; /* Maximum flow rate of device */

};
int main()
{

       FILE *fp;

        struct fplant s2;

       int i;
       int IDtoamend; /* User-entered ID variable */
       float fnewtemp;
       long int minusone = -1;
       int IDfound;

       /* Open tempflow.bin file */
       fp = fopen("tempflow.bin", "r");
       for (i = 0;i < 17;i++)
```

```
{
    /* Read each data structure from file sequentially */
    fread(&s2, sizeof(s2), 1, fp);

    /* Print each data structure */
    printf("\nID : %2d temp : %f flow rate : %f high temp : %f high
    flow : %f", s2.ID,s2.temp,s2.flowrate,s2.hightemp,s2.highflow);

}

fclose(fp);
/* User asked to enter the ID being monitored */
/* Go round "do loop" until a valid ID is entered */
IDfound=0;
do {
    fp = fopen("tempflow.bin", "r+");
    /* Ask user to enter ID */
    printf("\nenter ID   \n");

    scanf("%d", &IDtoamend);
    printf("\n ID is %d",IDtoamend);
    for (i = 0;i < 17;i++)
    {
        fread(&s2, sizeof(s2), 1, fp);

        if(IDtoamend == s2.ID)
        {
            /* Valid ID found */
            IDfound=1;
            break;

        }

    }
    if(IDfound==0)

        printf("\nID not found");
    fclose(fp);

} while( IDfound==0);
fp = fopen("tempflow.bin", "r+");
/* loop of 17 items in tempflow.bin file */
/* Need to find the user-entered ID */
```

```
        for (i = 0;i < 17;i++)
        {
            fread(&s2, sizeof(s2), 1, fp);

            if(IDtoamend == s2.ID)
            {
                /* Correct ID found in file */

                /* User asked to enter the new temperature being monitored */
                printf("\nenter new temperature   \n");

                scanf("%f", &fnewtemp);
                /* Print out confirmation of temperature to user */

                printf("\n new temperature is %f",fnewtemp);
                /* Store new temperature in file */

                s2.temp = fnewtemp;
                /* File updated with new temperature */
                /* As file pointer is currently pointing */
                /* to the next record in the file, we must */
                /* go back by 1 (minusone) to update the */
                /* correct record */
                fseek(fp,minusone*sizeof(s2),SEEK_CUR);

                fwrite(&s2, sizeof(s2), 1, fp);
                /* Print out the new values for the device */

                printf("\nID : %d temp : %f flow rate : %f high temp :
                %f high flow : %f", s2.ID,s2.temp,s2.flowrate,s2.hightemp,
                s2.highflow);
                /* Test if this new value is above the upper */
                /* limit for the temperature */
                if(s2.temp > s2.hightemp)
                    printf("\n ALERT! New temperature is above upper limit");

                break;

            }

        }

        fclose(fp);

    }
```

CHAPTER 8

1. Your program should be similar to this.

```
/* peke2.c */
/* Potential energy vs. kinetic energy */
/* also monitors total energy (KE + PE) */
/* */
#define _CRT_SECURE_NO_WARNINGS
#include <stdio.h>
#include <math.h>

#include <stdlib.h>
void main()

{
        int i;
        double m,g,t,h,hn,KE,PE;

        double u,v,totenergy,ttot;

        FILE *fptr;

        FILE *fptr2;
        fptr=fopen("peke.dat","w");
        fptr2=fopen("peke2.dat","w");
/* Set initial values from the formulas */
        m=10.0; /* Preset mass (kg) value */
        g=9.8; /* Preset acceleration of gravity (m/s2) value */
        h=10.0; /* Preset height (m) value */
        t=0.1; /* Preset time division (s) value */

        u=0.0; /* Preset initial velocity (m/s) value */
        totenergy=0.0;
        ttot=0.0; /* Set initial time value for total energy calculation */
        for(i=0;i<100;i++)
```

```
        {
            v=u+g*t; /* Find velocity v from initial velocity, accel. of
            gravity, and time */

            KE=0.5*m*pow(v,2); /* Find kinetic energy from mass and
            velocity */

            hn=u*t+0.5*g*pow(t,2); /* Find distance traveled in time t */

            h=h-hn; /* New height after falling hn meters */

            PE=m*g*h; /* Find potential energy */

            u=v; /* Set the initial velocity for the next increment of the
            loop to the current velocity */
    /* If h=0.0, then we have reached the ground */
            if(h<=0.0)
                break;
            totenergy=KE+PE; /* Find current total energy of the system
                                (kinetic + potential) */
            fprintf(fptr,"%lf\t%lf\n",KE,PE); /* Write current values to
                                                 KE v PE file */
            fprintf(fptr2,"%lf\t%lf\n",ttot,totenergy); /* Write current
                                                           values to KE + PE
                                                           file */
            ttot=ttot+75.0; /* Increment current time value */

        }
        fclose(fptr);

        fclose(fptr2);

    }
```

The corresponding graph for this code is shown in the following. It shows that the total energy (kinetic energy + potential energy) is constant.

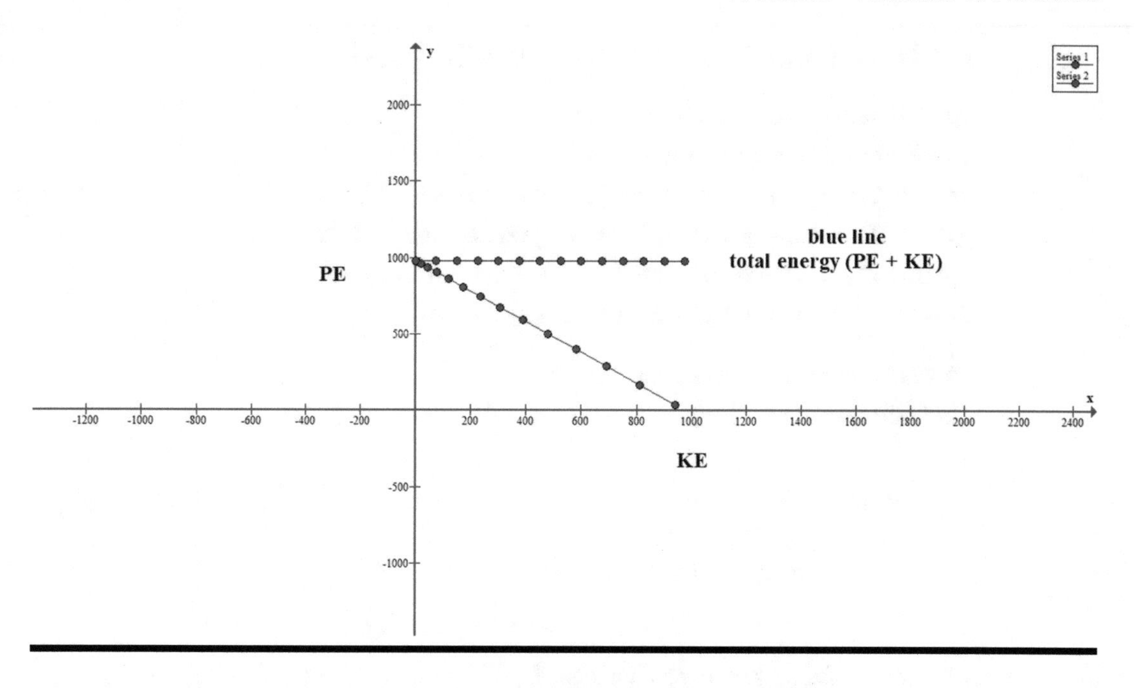

CHAPTER 9

1. The change to the original program was fairly small. Only the value of the length of the pendulum is changed. The code is shown in the following:

```
/* pendme3.c */
/*
 Euler-Cromer method
Changed length of cord
 */
#define _CRT_SECURE_NO_WARNINGS
#include <math.h>
#include <stdio.h>

void main()
{
       FILE *fptr;
       FILE *fptr2;
       int i,npoints;
```

```
    double length,g,dt,omega[250],theta[250],time[250];

    fptr=fopen("pendout3.dat","w");
    fptr2=fopen("pendout3b.dat","w");
    length=2.0; /* Preset length of pendulum (l) */
    g=9.8; /* Preset acceleration of gravity (m/s^2) */
    npoints=250; /* Preset number of points in loop */
    dt=0.04; /* Preset time interval (s) */

    /* Clear storage arrays to zero */
    for(i=0;i<npoints;i++)
    {
        omega[i]=0.0;
        theta[i]=0.0;
        time[i]=0.0;

    }
    /* preset theta and omega values */
    theta[0]=0.2;

    omega[0]=0.0;
    /* Euler-Cromer method */
    for(i=0;i<npoints;i++)
    {
        omega[i+1]=omega[i]-(g/length)*theta[i]*dt;
        theta[i+1]=theta[i]+omega[i+1]*dt;

        time[i+1]=time[i]+dt;
        fprintf(fptr,"%lf\t%lf\n",time[i+1],theta[i+1]);
        fprintf(fptr2,"%lf\t%lf\n",time[i+1],omega[i+1]);
    }
    fclose(fptr);
    fclose(fptr2);

}
```

Here is the output for the two omega (angular velocity) files.

The red graph is the original program with cord length 1m and the yellow graph is the new program with cord length 2m. You can see that there are two effects of changing the length of the cord.

i). The amplitude of the graph from the new program is less than that of the original. You could try an experiment with a pendulum yourself. You will notice that in the case of the longer pendulum, speed of the mass is not as high as with the shorter cord.

ii). The cycle of the pendulum for the longer cord is longer than for the shorter cord. This is shown by the lower frequency on the graph.

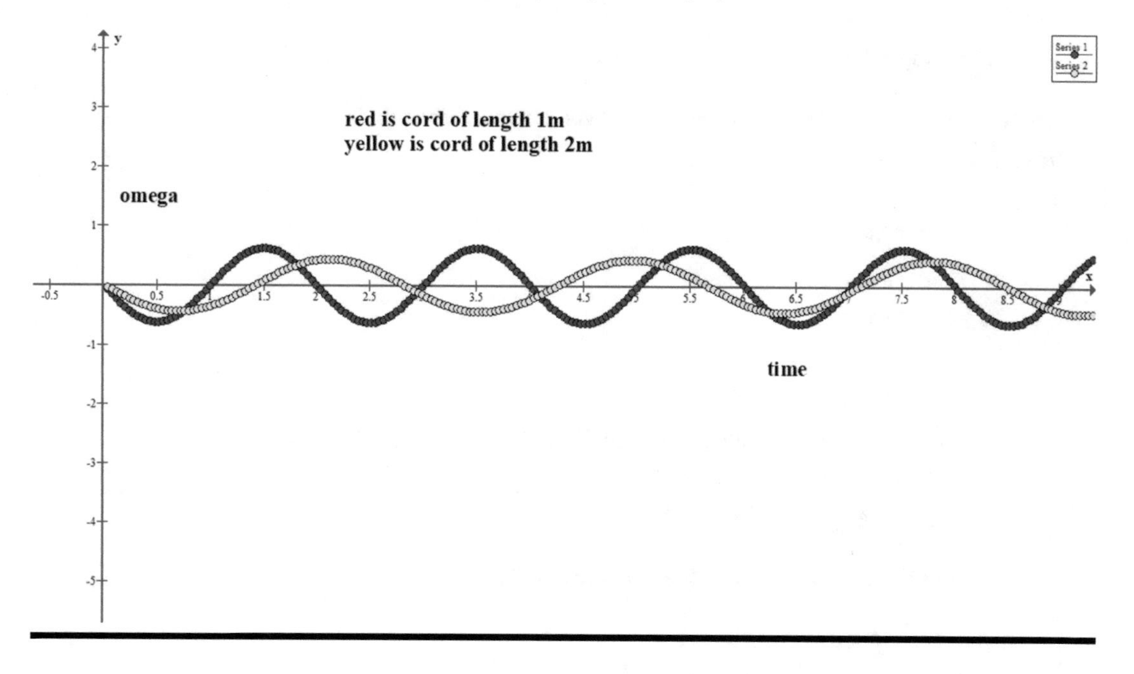

CHAPTER 10

1. The following code generates the graph shown:

```
/*      cofme.c
     Center of Mass Calculation.
       Calculates c of m for
       ellipse center = (0,0) a=2,b=1
*/
#define _CRT_SECURE_NO_WARNINGS
#include <stdlib.h>
#include <stdio.h>
#include <math.h>
#include <time.h>

double randfunc();/* Function to return random number */
void main()
{
       int  I,outcount;
       float area,total,count;
       FILE *fptr;
       time_t t;
       /* Local arrays */
      double x, y,xout[3500],yout[3500],xcofm,ycofm,a,b,root;
       double roottest1,roottest2;
       fptr=fopen("cofme.dat","w");

   /* Initializes random number generator */

   srand((unsigned) time(&t));
       /* clears arrays to zero */
         for( I = 0; I<3500;I++)
       {
            xout[I] = 0.0;
            yout[I] = 0.0;
       }
       /* Set x and y cofm accumulators to zero */
       xcofm=0.0;
       ycofm=0.0;
```

```
a=2.0;
b=1.0;
total = 0.0;
count = 0.0;
outcount = 0;
 for( I = 1;I<= 3500;I++)
{
/* Call random number function */
/* Get x values between -2 and +2 */
/* Get y values between -1 and +1 */
     x = randfunc()*4.0-2.0;

     y = randfunc()*2.0-1.0;
/* For the generated x values, if the y values are */
/* y > - sqrt((1/b^2)*(1-x^2/a^2)) and */
/* y < + sqrt((1/b^2)*(1-x^2/a^2)), then */
/* add 1 to count */
/* and update the x and y cofm values */
/* The preceding formulas are simplified to */
/* (1/b^2) is roottest1 and */
/* (1-x^2/a^2) is roottest2 */
roottest1=(1.0/pow(b,2));
roottest2=(1-(pow(x,2)/pow(a,2)));
root=sqrt(roottest1*roottest2);

     if(y>-root && y<root)
    {
     xcofm=xcofm+x;

     ycofm=ycofm+y;
     total = total+1;
     outcount = outcount +1;
     xout[outcount] = x;
     yout[outcount] = y;
    }

    count = count+1;

   }
```

```
        area=(total/count)*8; /* Area is part of rectangle which is 4x2
                                 or 8 sq units */

        printf("total is %f count is %f\n",total,count);

        xcofm=xcofm/total;
        ycofm=ycofm/total;

        printf("area is %lf\n",area);
        printf("cofm is %lf,%lf",xcofm,ycofm);
        /*  Plot the data */
        if(outcount >= 2700)

             outcount = 2700;
          for(I = 1; I<=outcount-1;I++)
             fprintf(fptr,"%lf %lf\n",xout[I],yout[I]);

         fclose(fptr);

}
double randfunc()
{
        /* Get a random number 0 to 1 */
        double ans;
        ans=rand()%1000;
        ans=ans/1000;
         return ans;

}
```

The resulting graph is shown in the following. The red dots are all of the accepted generated (x,y) points and the blue dot is at the center.

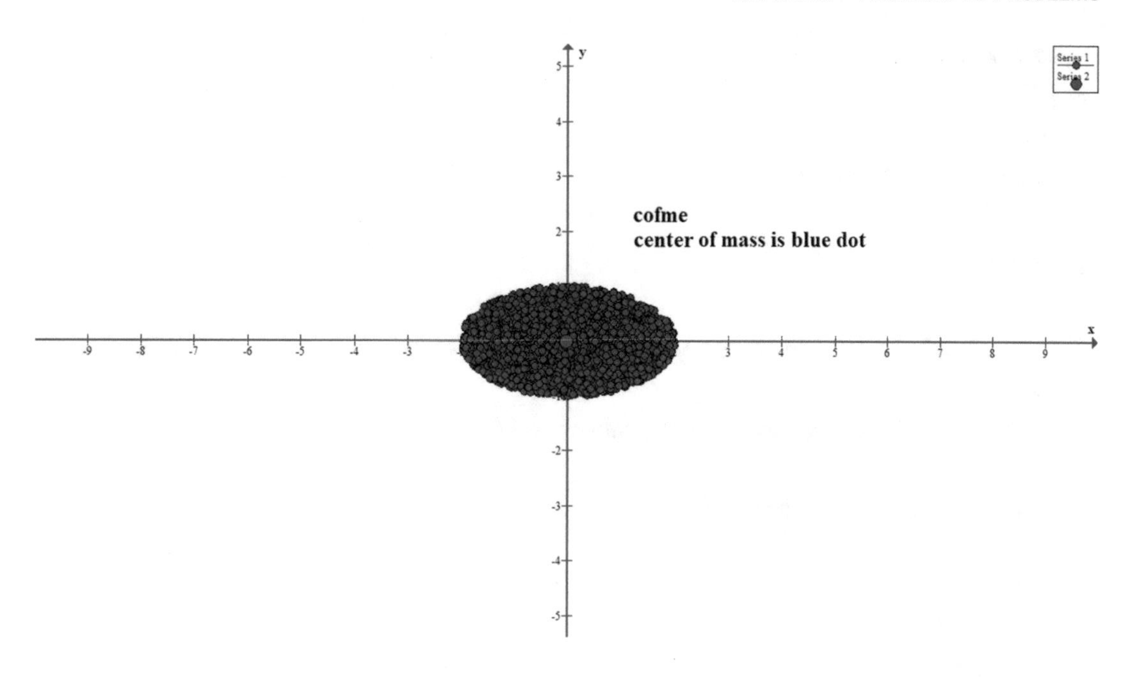

2. The following is the code which produces the graph:

```
/*      cofmcc.c
     Center of Mass Calculation.
Calculates c of m for
concentric circles
circle1 center = (0,0) radius = 2
circle2 center = (0,0) radius = 1
*/
#define _CRT_SECURE_NO_WARNINGS
#include <stdlib.h>
#include <stdio.h>
#include <math.h>
#include <time.h>
double randfunc(); /* Function to return random number */
```

```
void main()
{
int  I,outcount;
float area,total,count;
FILE *fptr;
time_t t;
/* Local arrays */

      double x, y,xout[3500],yout[3500],xcofm,ycofm;
fptr=fopen("cofmcc.dat","w");
   /* Initializes random number generator */
   srand((unsigned) time(&t));
/* clears arrays to zero */
   for( I = 0; I<3500;I++)
{
     xout[I] = 0.0;
     yout[I] = 0.0;
}
/* Set x and y cofm accumulators to zero */
xcofm=0.0;

ycofm=0.0;
total = 0.0;
count = 0.0;
outcount = 0;
 for( I = 1;I<= 3500;I++)
{
       /* Call random number function */
       /* Get x values between -2 and +2 */
       /* Get y values between -2 and +2 */
                x = randfunc()*4.0-2.0;

                y = randfunc()*4.0-2.0;
       /* For the generated x, the y values for the outer circle are  */
       /* y > -sqrt(4-pow(x,2) and */
       /* y < +sqrt(4-pow(x,2)) */
       /* and for the inner circle are */
       /* y > -sqrt(1-pow(x,2) and */
       /* y < +sqrt(1-pow(x,2)) */
       /* If this is true, then add 1 to count */
```

```
        /* and update the x and y cofm values */
            if(y>-sqrt(4-pow(x,2)) && y<sqrt(4-pow(x,2)))

           {
                 if(y>-sqrt(1-pow(x,2)) && y<sqrt(1-pow(x,2)))
               {
                 /* Exclude points inside inner circle */
                 goto out;
                 }
            xcofm=xcofm+x;

            ycofm=ycofm+y;
            total = total+1;
            outcount = outcount +1;
            xout[outcount] = x;
            yout[outcount] = y;

           }

out:            count = count+1;

   }
area=(total/count)*16; /* Area is part of the square which is 4x4
or 16 sq units */
printf("total is %f count is %f\n",total,count);
xcofm=xcofm/total;

ycofm=ycofm/total;
printf("area is %lf\n",area);

printf("cofm is %lf,%lf",xcofm,ycofm);

/*  Plot the data */
if(outcount >= 2700)

       outcount = 2700;
 for(I = 1; I<=outcount-1;I++)
       fprintf(fptr,"%lf %lf\n",xout[I],yout[I]);

fclose(fptr);

}
```

```
double randfunc()
{
        /* get a random number 0 to 1 */
        double ans;
        ans=rand()%1000;
        ans=ans/1000;
         return ans;

    }
```

The resulting graph is shown in the following. The red dots are all of the accepted generated (x,y) values. The blue dot is the center of mass. So the center of mass of a doughnut is in the middle of the hole.

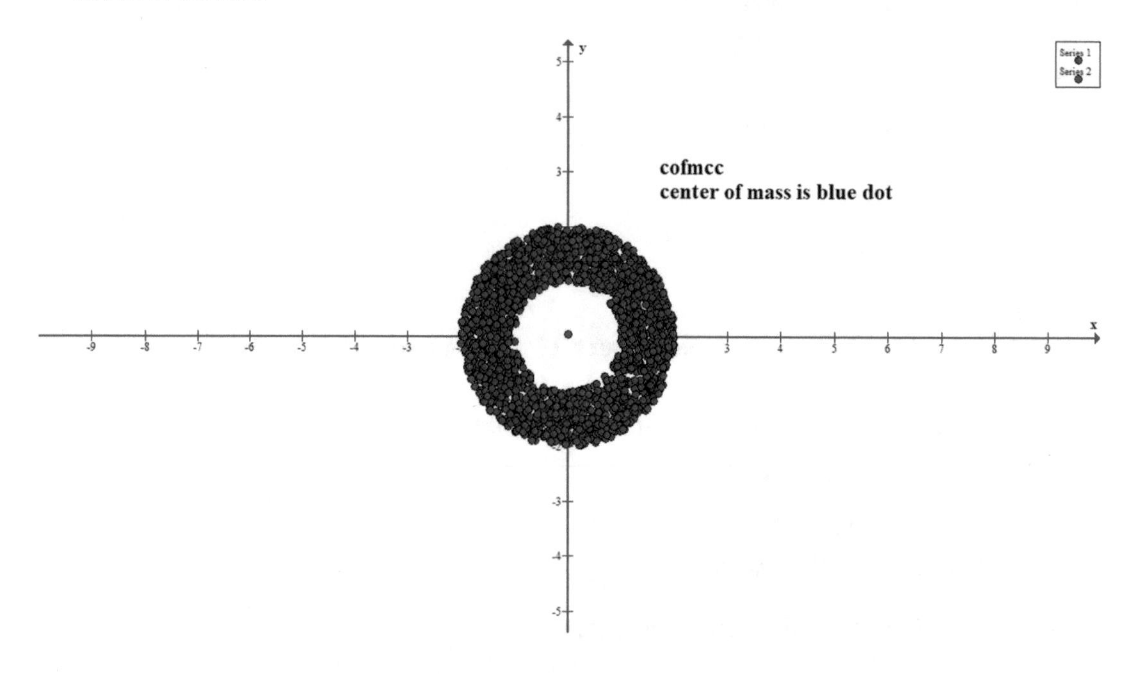

The program also prints out the area of the doughnut. If you run your program a few times, you will get slightly different values of this because of the different random numbers the program will generate. You can work out the area manually as shown as follows:

$$\text{Area} = \pi R^2 - \pi r^2$$

where **R** is the radius of the outer circle and **r** is the radius of the inner circle.

So for the two radii of 2 and 1 units, we get the area to be 9.4247779.

If you run the program a few times and take the average of the area values printed by the program, then you should get a value to within two decimal places of the previous manually calculated value.

CHAPTER 11

1. The following is an example of the program:

```
/* Brownian motion (2D) Simulation (Monte Carlo)*/
/* selects x and y changes */
#define _CRT_SECURE_NO_WARNINGS
#include <stdlib.h>
#include <stdio.h>
#include <math.h>
#include <time.h>

void main()
{
       FILE *fptr;

       time_t  t;
int i;
       int collisions;
       double xrand;
       double yrand;
       double xplusminusrand;
       double yplusminusrand;
       double xvals[5950],yvals[5950];
       double cosval,sinval;

       fptr=fopen("browntxy.dat","w");

/* Set the random number seed */

       srand((unsigned) time(&t));
       collisions=1000;
       xvals[0]= 0.0;
```

```
yvals[0]=0.0;
for(i=0;i<1000;i++)
{
    /* Random x value 0-1 */
    xrand=rand()%1000;
    xrand=xrand/1000;

    /* Random y value 0-1 */
    yrand=1.0-xrand; /* random y value */
    xplusminusrand=rand()%1000;
    xplusminusrand=xplusminusrand/1000;

    /* Randomly find + or - */
    /* As our random number is */
    /* between 0 and 1, we can take */
    /* any number less than 0.5 to be */
    /* minus in our calculation and */
    /* numbers above 0.5 to be plus */
    if(xplusminusrand < 0.5)
        xplusminusrand = -1.0;
    else

        xplusminusrand = 1.0;
    yplusminusrand=rand()%1000;
    yplusminusrand=yplusminusrand/1000;

    /* Randomly find + or - */
    if(yplusminusrand < 0.5)
        yplusminusrand = -1.0;
    else

        yplusminusrand = 1.0;
    /* Move particle by x amount */
    xvals[i+1]=xvals[i]+xrand*xplusminusrand;
    cosval = xrand*xplusminusrand; /* possible equivalent cos value */

    /* Move particle by y amount */
    yvals[i+1]=yvals[i]+yrand*yplusminusrand;
```

```
            sinval = yrand*yplusminusrand;/* possible equivalent sin value */
            fprintf(fptr,"%lf %lf\n", xvals[i], yvals[i]);

            /*printf("cosval = %lf sinval = %lf\n",cosval,sinval);*/

        }

}
```

And this is a possible output.

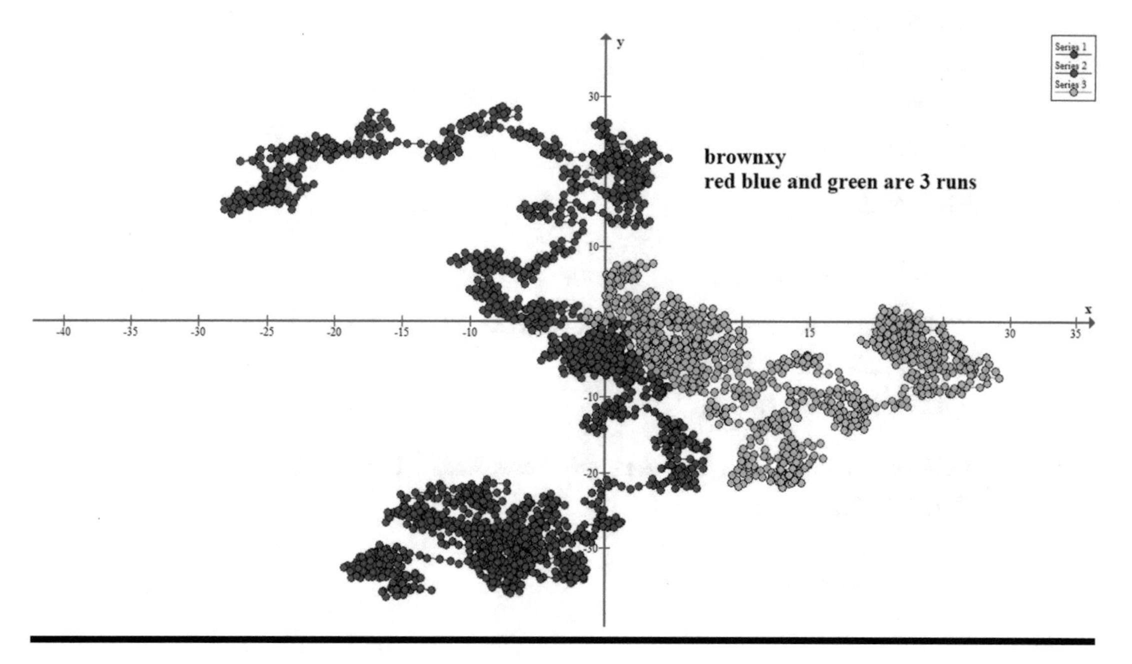

CHAPTER 12

1. The program is shown in the following:

```
/*      PROGRAM vacxaf.c
        VACANCY DIFFUSION MODEL. (2D VERSION)
*/
#define _CRT_SECURE_NO_WARNINGS
#include <stdlib.h>
#include <stdio.h>
#include <math.h>
```

```
#include <time.h>

int IRND();/* Function to return random number 0 to 19 */
int IFOURRND();/* Function to return random number 1, 2, 3, or 4 */

void main()

{
        int N1,N2;
        int N1N,N2N,MCC;
        int LATTICE2[20][20];
        int MCCMAX;
        int Q,P,INC;

        FILE *fptr;

        time_t t;

        /* Initializes random number generator */
        srand((unsigned) time(&t));
        fptr=fopen("vacxaf.dat","w");

        MCCMAX = 1000; /* Set number of Monte Carlo Cycles */
        for(P=0;P<20;P++)
        {
             for(Q=0;Q<20;Q++)
             {

             /* FILL THE ARRAY */

                   LATTICE2[P][Q] = 0;
             }

        }
        /* SELECT ANY SITE AS THE INITIAL VACANCY SITE*/
        /* Can be set randomly using the IRND function */
        /* or can be set to specific values */

        /*N1N = IRND();
        N2N = IRND();*/
        N1N=1; /* Start x value */
        N2N=10; /* Start y value */
```

```
LATTICE2[N1N][N2N] = 1; /* Set vacancy site in lattice */

for(MCC=1;MCC<=MCCMAX;MCC++)

{
     N1=N1N;
     N2=N2N;

     if(LATTICE2[N1][N2] == 1)
     {

            /* VACANCY SITE (= 1 )*/

            INC = IFOURRND(); /* Call function to randomly select 1, 2,
                                 3, or 4 */
            /* 1 indicates a move to the right */
            /* 2 indicates a move up */
            /* 3 indicates a move to the left */
            /* 4 indicates a move down */
            /* Instead of going from 19 to 1, etc., you bounce off the
            boundary. Go from 19 to 18, etc. */
            if(INC == 1) /* right */
            {
                   if(N1 == 9)
                   {
                           if (N2 == 9 )
                                N1N=N1N+1;
                          else
                                N1N=N1N-1;
                   }
                   else
                     if(N1 == 19)
                          N1N = 18;
                   else

                          N1N = N1+1;
                   }else if(INC == 2) /* Up */
                   {
                           if(N2 == 19)
                                N2N = 18;
```

```
                        else

                             N2N = N2+1;
                  }else if(INC == 3) /* Left */
                  {
                        if(N1 == 10)
                        {
                              if (N2 == 9)
                                   N1N=N1N-1;
                             else
                                   N1N=N1N+1;
                        }
                        else
                          if(N1 == 1)
                             N1N = 2;
                        else

                             N1N = N1-1;
                  }else if(INC == 4) /* Down */
                  {
                          if(N2 == 1)
                             N2N = 2;
                        else

                             N2N = N2-1;

                  }
                  if(LATTICE2[N1N][N2N] == 0)
                  {

                        LATTICE2[N1N][N2N] = 1; /* Set as a used site */
                  }
                  else
N1N=N1N;
                        /*printf("not found\n");*/

            }

      }
      /* Write any used lattice positions to file */
```

```
        for(P=0;P<20;P++)
        {
              for(Q=0;Q<20;Q++)

              {
                     if(LATTICE2[P][Q] == 1)

                            fprintf(fptr," %d\t%d\n",P,Q);
              }

        }

        fclose(fptr);

}
int IRND()
{
        /* Generate a random whole number from 0 to 19 */
        double TOT,DIV,X;

        int ANS,I;
        TOT=rand()%1000;

        TOT=TOT/1000;
        /* Returns 0, 1, 2, … or 19 */
        /* chosen at random */

        DIV = 20.0;
        X = 1.0;
        for(I=0;I<20;I++)
              if(TOT < X/DIV)

                     ANS = I;
              else

                     X = X+1.0;
        return ANS;

}
```

```
int IFOURRND()
{
        /* Generate a random whole number 1, 2, 3, or 4 */
        double TOT;

        int ANS;
        TOT=rand()%1000;
        TOT=TOT/1000;

        /* Returns 1, 2, 3, or 4 */
        /* chosen at random */

        if(TOT < 0.25)
             ANS = 1;
        else if(TOT < 0.5)
             ANS = 2;
        else if(TOT < 0.75)
             ANS = 3;
        else

             ANS = 4;

        return ANS;
}
```

The associated graph is shown in the following. Your graph will probably be slightly different to this one as the random numbers generated will be different each time the program is run. As the "wall" is at x=9 and the "hole" is at (9,9), then you can see where the "particle" has crossed. Because the particle started at (1,10), then it looks as if it moved around the right side for a short time, and then it crossed to the left and then was constricted to the left side – although it may have crossed back and forth!

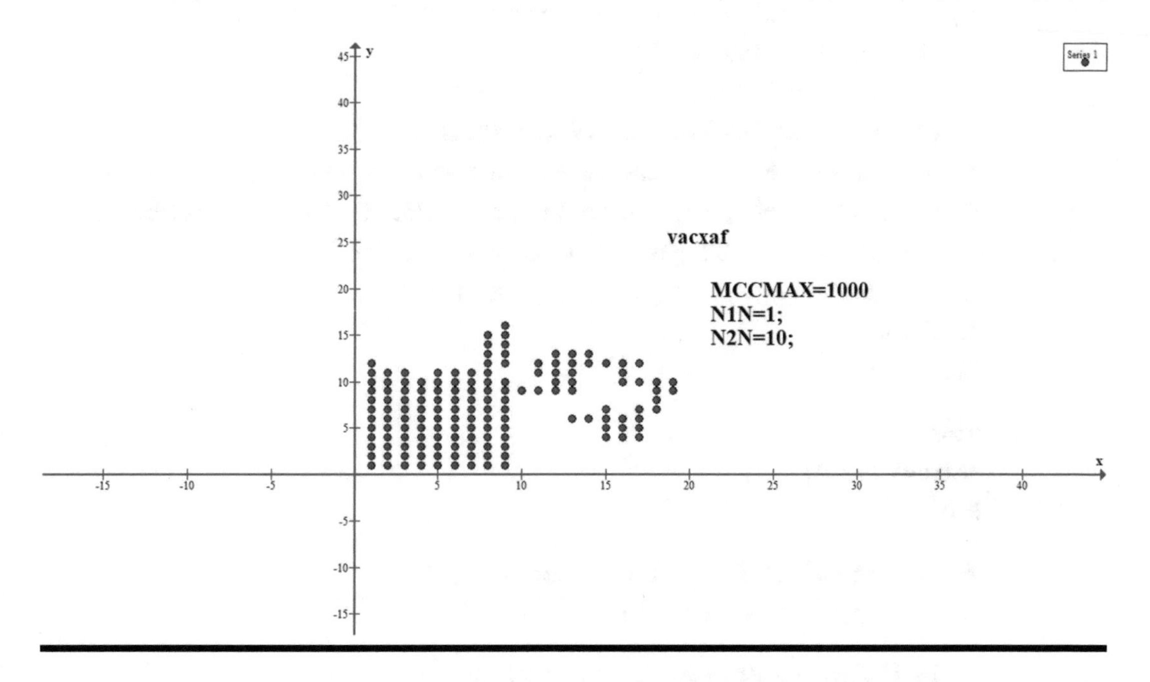

CHAPTER 13

1.

```
/*      chresp.c
     Chain Reaction Simulation.
        Volume of sphere */
#define _CRT_SECURE_NO_WARNINGS
#include <stdlib.h>
#include <stdio.h>
#include <math.h>
#include <time.h>

double randfunc(int N); /* Function to return random number (0 to 1)*/

int checkin(double x,double y,double z,double r); /* Function to check if the
                                                     particle dimensions are
                                                     inside the sphere */
void main()
{
       FILE *fptr;
       FILE *fptr2;
```

```
   /*   Local variables  */
      int   K,P,N,Ninp;
/* x0, y0, z0 is the position of the fission nucleus */
/* x1, y1, z1, phi1, d1, costheta1 are positions of first neutron */
/* x2, y2, z2, phi2, d2, costheta2 are positions of second neutron */
double f,x0,y0,z0,phi1,phi2,d1,d2,costheta1,costheta2;
double Vinit,Vfin,W,r,x1,y1,z1,x2,y2,z2,third;
double pi;
time_t t;

pi=3.142;
third=1.0/3.0;
P=0;

/*  Select output file for error messages */
fptr = fopen("chresp.err","w");

/* Initialize random number generator */
srand((unsigned) time(&t));

/* Ask the user for the number of fissions */
printf("Enter number of fissions \n");
scanf("%d",&N);

/* Create results file */
fptr2 = fopen("chresp.dat","w");
if( fptr2 == NULL)
{
fprintf(stderr,"Error writing to %s\n","chresp.dat");
fclose(stderr);
return(1);

}

/* Initial and final values of volume of sphere */
Vinit=0.0001;
Vfin=1.0;

/* forloop for initial volume Vinit, */
/* final volume Vfin, and increments */
```

```
/* of 0.0005*/

for(W=Vinit;W<Vfin;W=W+0.0005)
{
     r=pow((3*W)/(4*pi),third);
     Ninp = 0;
     for(K=1;K<=N;K++)
     {
            /* Find a random position within the sphere */
            /* for the nucleus */
            x0 = r*(randfunc(P)-0.5);
            y0 = r*(randfunc(P)-0.5);
            z0 = r*(randfunc(P)-0.5);
            phi1 = 2*pi*randfunc(P);
            costheta1 = 2*(randfunc(P)-0.5);
            phi2 = 2*pi*randfunc(P);
            costheta2 = 2*(randfunc(P)-0.5);
            d1 = randfunc(P);
            d2 = randfunc(P);

            /* Calculate the position of first neutron */
            x1 = x0 + d1*sin(acos(costheta1))*cos(phi1);
            y1 = y0 + d1*sin(acos(costheta1))*sin(phi1);
            z1 = z0 + d1*costheta1;

            /* Calculate the position of second neutron */
            x2 = x0 + d2*sin(acos(costheta2))*cos(phi2);
            y2 = y0 + d2*sin(acos(costheta2))*sin(phi2);
            z2 = z0 + d2*costheta2;

            /* Find out if first neutron is inside the sphere */
            if(checkin(x1,y1,z1,r) == 1)
              Ninp = Ninp+1;

            /* Find out if second neutron is inside the sphere */
            if(checkin(x2,y2,z2,r) == 1)
              Ninp = Ninp+1;
        }
        f = (double)Ninp/(double)N;
```

```
            fprintf(fptr2,"%lf %lf\n",r,f);
        }
        fclose(fptr2);
        fclose(stderr);
}
double randfunc(int N)
{
        /* Find a random number between 0 and 1 */
        double TOT;
        TOT=rand()%1000;
        TOT=TOT/1000;
        return TOT;

}
int checkin(double x,double y,double z,double r)
{
        /* If the coordinates are within the sphere, return 1 */
        /* Otherwise return 0 */
        int I;
        if(x < r && y < r && z < r)
            I = 1;
        else
            I = 0;
            return I;

}
```

The resulting graph is shown in the following.

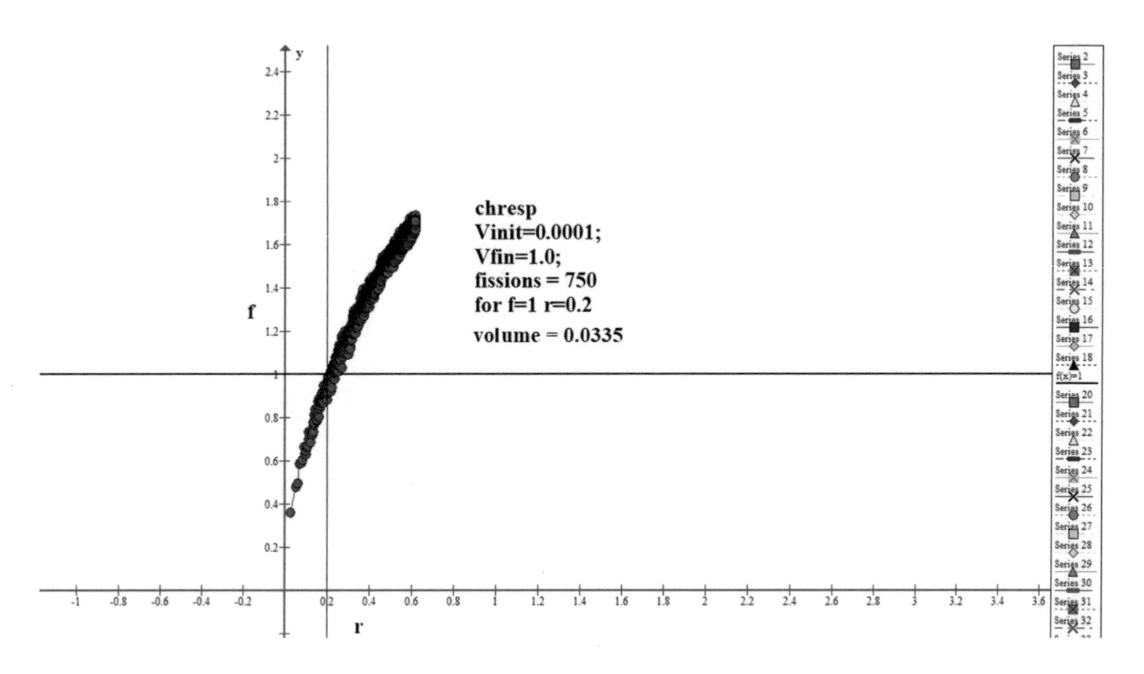

The value of r where f = 1 can be used to find the volume of the sphere. If we compare this with the volume of our cube, we will see that we need a much smaller volume for the sphere to reach critical mass than we had for the volume of the cube.

Index

P. Joyce, *Practical Numerical C Programming*, https://doi.org/10.1007/978-1-4842-6128-6

F

G, H

I, J, K

L

M

N, O

P, Q